W9-AED-914

DISCARDED
JENKS LRC
GORDON COLLEGE

Pergamon Series in Analytical Chemistry
Volume 7

General Editors: R. Belcher† (Chairman), D. Betteridge & L. Meites

The Analysis of
Gases by Chromatography

Related Pergamon Titles of Interest

Full details of any Pergamon publication and a free specimen copy of any Pergamon journal available on request from your nearest Pergamon office.

The Analysis of Gases by Chromatography

by

C. J. Cowper and A. J. DeRose

British Gas Corporation
London Research Station

PERGAMON PRESS

OXFORD · NEW YORK · TORONTO · SYDNEY · PARIS · FRANKFURT

WINN LIBRARY
Gordon College
Wenham, Mass. 01984

U.K.	Pergamon Press Ltd., Headington Hill Hall, Oxford OX3 0BW, England
U.S.A.	Pergamon Press Inc., Maxwell House, Fairview Park, Elmsford, New York 10523, U.S.A.
CANADA	Pergamon Press Canada Ltd., Suite 104, 150 Consumers Road, Willowdale, Ontario M2J 1P9, Canada
AUSTRALIA	Pergamon Press (Aust.) Pty. Ltd., P.O. Box 544, Potts Point, N.S.W. 2011, Australia
FRANCE	Pergamon Press SARL, 24 rue des Ecoles, 75240 Paris, Cedex 05, France
FEDERAL REPUBLIC OF GERMANY	Pergamon Press GmbH, Hammerweg 6, D-6242 Kronberg-Taunus, Federal Republic of Germany

Copyright © 1983 C. J. Cowper and A. J. DeRose

All Rights Reserved. No part of this publication may be reproduced, stored in a retrieval system or transmitted in any form or by any means: electronic, electrostatic, magnetic tape, mechanical, photocopying, recording or otherwise, without permission in writing from the publishers.

First edition 1983
Reprinted 1984

Library of Congress Cataloging in Publication Data
Cowper, C. J.
The analysis of gases by chromatography.
(Pergamon series in analytical chemistry; v. 7)
Bibliography: p. 131
Includes indexes.
1. Gases—Analysis. 2. Gas chromatography.
I. DeRose, A. J. II. Title. III. Series.
QD121.C67   1983   543'.0897   83-6207

British Library Cataloguing in Publication Data
Cowper, C.J.
The analysis of gases by chromatography. —
(Pergamon series in analytical chemistry; v.7)
1. Gas chromatography
I. Title II. DeRose, A.J.
544'.926   QD79.C45
ISBN 0-08-024027-5

In order to make this volume available as economically and as rapidly as possible the authors' typescripts have been reproduced in their original forms. This method unfortunately has its typographical limitations but it is hoped that they in no way distract the reader.

Printed in Great Britain by A. Wheaton & Co. Ltd., Exeter

QD
121
.C67
1983

Foreword

Professional and practising analysts come of a long and highly regarded lineage. But amongst them gas analysts have undoubtedly been the poor relation. Their science and their art have been neglected by both academic and industrial researchers intent on using up-to-date techniques, large machines and sophisticated instrumentation. The 19th and early 20th century workers - Hempel, Orsat, Haldane and others less familiar who left their names associated with particular apparatus - seemed to have few if any successors in the towns' gasworks laboratories where the tradition of gas analysis were for so long preserved.

As so often happens, the need for change coincided with the means for undertaking it. The demand of the sixties and seventies for more, more accurate and more rapid analysis of highly complex gaseous mixtures has been well met with this application of gas chromatography. For it a whole new branch of industrial analytical technology has been created and then applied in both old and new industries as well as in research laboratories where gaseous materials have assumed a new perspective and importance.

Our own and earlier modest contribution to the recording of this new technology - "Gas analysis by gas chromatography" - was enthusiastically received. The same enthusiasm will, I am sure, greet this new book which I welcome as a much needed and long overdue contribution to the analytical science of the eighties.

P.G. Jeffery
Laboratory of the Government Chemist

Preface

Gases are of great importance in industry as raw material, fuels, primary products, by-products and wastes. They have great environmental significance since we breathe gas mixtures every minute of our lives. Gas analysis is important for quality control, process control, evaluation of process efficiency, safety, environmental monitoring and fundamental physico-chemical studies.

The purpose of this book is to assist scientists and technicians who have a basic knowledge of chemistry and physics and an appreciation of the principles and practice of gas chromatography to select equipment for and to perform the analysis of complex gas mixtures.

Although in Chapter 1 we identify certain specific requirements for analysis of gases rather than liquids, there are more similarities than there are differences between the areas of application. In the same way, it would be foolish to make a dogmatic limitation to the scope of the book by only considering those components whose boiling points are below ambient temperature, or more rigorously whose critical temperatures are below ambient. Thus hexane, cyclohexane, benzene and many other hydrocarbons are components of natural gas, and we consider them in this context. Gases dissolved in liquids are included, and also materials such as propane and halocarbons which are liquefied under pressure. Specific areas of application such as atmospheric pollution and head space analysis are not included since they have been dealt with comprehensively elsewhere, though in areas common to these and other applications the book may give useful information.

Since Jeffery and Kipping revised their very valuable book "Gas Analysis by Gas Chromatography", there have been significant changes in the subject. In that book, and in Leibrand's (1967) very comprehensive Atlas of Gas Analysis, most applications are illustrated by isothermal, single-column chromatograms. More recently, many gas analyses have been performed on multi-column systems, with the groups of components being routed onto appropriate columns by valves or pressure-balancing mechanisms. Many multi-column methods are described in Thompsons's (1977) book, which also illustrates, via its chromatograms, the improved separation efficiency which gas analysts can now obtain with modern equipment.   The apparatus has become more complex and comprehensive (few analysts now make their own), and the impact of microprocessor systems on control of instruments and on data handling has blurred the distinction between laboratory and process chromatographs.

We have dealt with some aspects of equipment in detail, either because they are particularly relevant in the context of gas analysis, because they have not been comprehensively treated in the literature or are not given the attention they deserve by manufacturers.   Thus detectors are described at some length, because the gas analyst is more likely than the liquid analyst to encounter a range of problems which require a variety of detectors, and a thorough understanding of their operation aids the choice.   The differences between alternative methods of gas flow control and measurement are often not fully appreciated, and so we give considerable detail.   By the same token, we do not discuss methods of temperature control or signal handling, since virtually all manufacturers deal very adequately with these aspects.

We are grateful to the British Gas Corporation for permitting us to prepare and publish this book.   We thank our numerous colleagues at London Research Station who have assisted in its preparation, particularly Marie Brennan, and our numerous contacts in the British gas industry, the European gas industry and fellow members of various technical and standardisation committees who have contributed to the information recorded here but could not be individually acknowledged.   We also thank Carle Instruments Inc, Hewlett-Packard Ltd, Perkin Elmer Ltd, Pye Unicam Ltd, Tracor Inc., and Varian Associates Ltd for supplying and allowing us to publish drawings of their products.

Contents

AGC-A*

CHAPTER 1

Introduction

1.1 HISTORY

Gas chromatography was first demonstrated for analysis of liquids in the classic paper by James and Martin (1952). It was quickly recognised as a rapid and quantitative method for solving hitherto intractable problems, and within a few years became the most popular instrumental method of analysis. The original explosive stimulus to growth came from the petrochemical industry, but it was soon adopted in every area where mixtures of organic (and sometimes inorganic) materials were required to be analysed. Its growth in both number and range of applications is reflected in the abstracts produced by the Chromatography Discussion Group, the Preston Technical abstracts and the different series of International Symposia. The developments in theory and instrumentation which attended this growth provided a basis for the more recent development of high performance liquid chromatography.

The fact that chromatography is a separation technique provoked comparison with analytical distillation and hence the theoretical plate concept was introduced to describe performance. Chromatography columns are easily able to achieve 100 times more plates than distillation columns of comparable length, and are capable of generating 10,000 times more plates. This performance, used with stationary phases of different polarities and selective detectors provides a very powerful analytical tool.

Although James and Martin's paper is generally regarded as the starting point, the first reference to gas analysis by gas chromatography was made by Prior (1947) five years earlier. Gas analysis was, however, overwhelmed by the application to analysis of liquids which followed the James and Martin paper. Although gas analysis methods undoubtedly benefitted from the general expansion in gas chromatography, it is important to recognise that gas analysis was for many years neglected by instrument manufacturers, by theorists and by the literature (although Jeffery and Kipping (1964, 1972) did much to redress the balance). Manufacturers have more recently produced a number of application notes on gas analysis, but it is still represented as a black art.

1.2   DIFFERENCES BETWEEN GAS AND LIQUID ANALYSIS

1.2.1   Equipment and Materials

The main market for gas chromatographs is for liquid analysis, and consequently the "average chromatograph" will have a heated liquid injection
device, a column oven designed for operation at temperatures substantially
above ambient, and a flame ionisation detector.   By contrast the "average
instrument" for gas analysis will have a gas sampling valve which for most
purposes need not be heated, a column oven which can operate close to or
below ambient temperature, and a thermal conductivity detector.   If a
septum injector is fitted, an unheated one would be preferred, for simplicity and low dead volume, and the instrument should be capable of accep-
ting switching valves or other such devices.   Apparatus is more fully
described in Chapter 2, and we would only state here that a design which is
satisfactory for liquid samples may not always be successfully adapted for
gas analysis.

Stationary phases are also very different.   When analysing liquids, as a
general rule of thumb the phase may be chosen to be chemically similar to
the sample, and operated at a temperature or over a temperature range which
is similar to the boiling point or range of the sample.   Neither of these
ideas are practicable when the sample is a gas, and hence separation of
gases has largely been achieved by using adsorption rather than partition.
Adsorption columns operate at temperatures substantially higher than the
boiling points of the samples for which they are used, and consequently can
be used within the normal temperature range of chromatographic ovens.   Thus
helium (B.Pt $-269^{o}C$), hydrogen ($-253^{o}C$), oxygen ($-183^{o}C$) and nitrogen
($-196^{o}C$) are conveniently separated on molecular sieve at $50^{o}C$.   Chapter 3 describes the properties of the various stationary phases used.

1.2.2   Theory

Most models of gas chromatography treat the carrier gas merely as a vehicle
for the sample components during the time they spend in the vapour phase.
Gas-phase interactions can occur such that a change of carrier gas can
change the order of elution of a pair of components (Cremer, reported by
Keulemans, 1959), but they are fortunately rare.   In gas analysis, the
carrier gas has physical properties which are similar to those of sample
components, indeed it may even be a sample component, and so cannot be
treated simply as an inert vehicle.   The concept of gas hold-up time as a
measure of carrier gas velocity becomes difficult to apply.   Measurement of
an "air peak" is meaningless when air can be separated into its components.

Column performance can be affected by the choice of carrier gas, as the
adsorbent surfaces become fully saturated with the carrier.   Chapter 4
describes the choice of carrier gas, and demonstrates how two components may
no longer be separated once they are both present in the carrier.   Adsorbents such as molecular sieve, silica gel and alumina can also adsorb
moisture from the atmosphere, from carrier gas or from samples and consequently change their characteristics.

For the above reasons, conventional means of expressing elution order, such
as adjusted relative retention times or retention indices should be used
with great caution, and it is probably this fact which gives the black art
image referred to earlier.   "Spiking" of samples with known components,

comparison of response between different detectors and different carrier gases and a knowledge of the physical chemistry of the separation process are all valuable aids to identification.

1.2.3   Techniques and Tactics

Many problems in gas analysis cannot be solved by the conventional methods of performing the analysis at a different temperature or using a column of higher efficiency.   The best known example concerns gases containing oxygen, nitrogen and carbon dioxide.   Separation of O_2 and N_2 without using sub-ambient temperatures requires a molecular sieve column, on which CO_2 has such a long retention time that it is often considered to be irreversibly adsorbed.   Many columns are available which separate CO_2 from air, but none will at the same time separate O_2 from N_2.   A number of techniques for column switching, backflushing etc have been developed for and widely applied in gas analysis.   These tactical approaches to separation problems are described in Chapter 5.

1.3   PROCESS GAS ANALYSIS

Process applications have evolved somewhat separately from the main body of gas chromatography, but recent instrumental developments, particularly microprocessor control, have caused process and analytical techniques to converge.   In this book we have not attempted to describe the specialised hardware required for process chromatographs but we would expect the separation tactics described later to be useful to the engineer who is considering process problems as well as to the laboratory chemist.

1.4   IDEAL AND NON-IDEAL BEHAVIOUR

Ideal gases, as defined by Avogadro's Law, behave in such a way that volumes are always additive and can be unequivocally translated to molar or mass quantities.   Unfortunately most gases behave non-ideally to a greater or lesser extent, and so it is an approximation to assume that volumes of different gases are additive, or that equal volumes contain the same number of molecules.   This behaviour is caused by molecular interactions and finite molecular size, and is normally characterised by the compressibility factor Z, where

$$Z = \frac{PV}{nRT}$$

The compressibility factor, which equals 1 for an ideal gas under all conditions, is a function of both temperature and pressure, but the values of interest in gas analysis are those which obtain at ambient conditions. Other than for helium and hydrogen, all these values are less than 1, and so a given volume of real gas contains more molecules than if it were ideal. Some typical values are given in Table 1.1

Mixtures of gases have compressibility factors which are not simply the sum of the partial contributions of their components, but can be calculated. For fuel gas mixtures, the method of Mason and Eakin (1961) can be used.

Non-ideal behaviour affects the gas analyst in two ways.   The first concerns the preparation of standard gases in cases where the composition

Table 1.1   Compressibility Factor of Pure
Gases at 15°C, 1 atmosphere

Gas	Z
Hydrogen	1.0006
Oxygen	0.9993
Methane	0.9981
Hydrogen sulphide	0.9903
Propene	0.9843
Sulphur dioxide	0.9790
n-Butane	0.9639

is calculated from physical measurements which are influenced by non-ideal behaviour. This is described in Chapter 9. The second concerns the effect on injected sample size, since injection devices define a volume and not a molar or mass quantity. Compressibility factors of binary mixtures approximate to the values for the major components, so that a mixture of exactly 1% molar CH_4 in H_2 would have Z = 1.0006, and exactly 1% molar CH_4 in n-butane would have Z = 0.9639. If one of these mixtures were used as an external standard for measurement of the CH_4 content of the other, there would be a 4% relative difference in the amount of CH_4 injected. This is an extreme example, and in many cases the error may not be significant, but it illustrates how, for the most accurate analysis, the composition of standard and sample should be close to each other, not only for analysed components but also for the matrix gas.

1.5   UNITS

Throughout the book a variety of practical units have been used. While it is desirable to have a single, consistent set of units, the real situation prevents this, and some apparently bizarre combinations of units are used, for perfectly good reasons. For example, those dimensions of a column which are significant to the separation i.e length and inside diameter, are expressed in metric units (m and mm respectively). The dimensions which show whether in practical terms a column can be connected into a chromatograph, e.g. outside diameter, are almost invariably expressed in Imperial units (1/4 in. or 1/8 in.).

The units used include:

Length: Metres, millimetres, micrometres, inches and feet.

Volume: Cubic metres, litres, millilitres and microlitres.

Temperature: Degrees Celsius.

Pressure: Pascals (newtons/m^2), pounds per square inch, and atmospheres.

Composition: The preferred units are mole percent (number of molecules of component per 100 molecules of sample) or parts per million molar (number of molecules of component per 10^6 molecules of sample).

Composition expressed on a mass basis is less readily

appreciated, but translates readily into molar terms. Volume percent is ambiguous unless further defined; unfortunately many gas analysts have a much clearer understanding of volume composition as a result of methods of preparing gas mixtures and of analysers such as the Orsat apparatus than of molar or mass composition.

Mesh size: ASTM sieve size and particle size in micrometres are used as appropriate.   The relationship between them and British Standard mesh sizes is given in Table 1.2.

Table 1.2   Particle and Mesh Sizes

Particle diameter micrometres	ASTM mesh (ASTM E11-61)	BS mesh (BS 410)
595	30	25
420	40	36
355	45	44
250	60	60
177	80	85
149	100	100
125	120	120
105	140	150

1.6   DRAWING CONVENTIONS

There is no uniformity in the literature on how gas chromatographic systems are diagramatically represented.   In this book labelled boxes (e.g. for columns), miniature drawings (e.g. for double oblique bore taps), engineering line drawing conventions (pressure regulators, etc.) and electrical circuit conventions (gas flow restrictors ≡ resistors) have been used as appropriate.   Liberal word labelling should assist those unfamiliar with the conventions used.

CHAPTER 2

Equipment

2.1 INTRODUCTION

The main market for gas chromatographic equipment is for liquid analysis and most, though not all chromatographs are designed primarily for this market. Most manufacturers sell adaptations of their models for gas analysis and these adaptations and needs are especially considered in this Chapter. Excellent, though dated, advice on the selection of equipment for gas analysis is given in British Standard BS 4857 (1970).

Different means of sample introduction are used in gas analysis, and some detectors are particularly well suited to this need.   For this reason these two sections are more detailed than others.   Gas flow controls are also sometimes misused through misunderstanding, and so their mechanism and use are described at some length.

2.2 GAS CONTROLS

Control of gas flows is necessary for repeatable separation and the effective and stable operation of detectors.   Control is achieved by the use of pressure regulators and flow controllers, and means of measuring pressure and flow allow the effectiveness of the control to be estimated.

2.2.1   Pressure Regulators

The pressure regulator is an attractively simple device, producing an adjustable outlet pressure which is substantially independent of the pressure of the source from which it is fed.   The flow of gas through the regulator depends upon the down-stream resistance and, at any particular pressure setting, can vary over the range likely to be used in gas chromatography without effect on the outlet pressure.   Figure 2.1 is a schematic drawing of a pressure regulator.

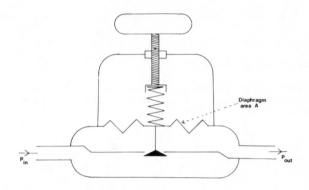

Fig. 2.1.   Pressure regulator

Gas enters the lower chamber, and passes via the valve to the upper chamber and thence to the outlet.   The position of the valve is dictated by the forces acting on the diaphragm, the spring pressure pushing downwards, and the outlet pressure pushing upwards.   If the outlet pressure rises above the set value the valve closes and the flow of gas stops.   If the pressure drops below the set value, the valve opens, and the downstream volume is rapidly repressurised.

In fact, an error term is introduced by the fact that the pressures on either side of the valve are unequal.   The balance of forces in the regulator is as follows.   Downward pressure is the sum of the force exerted by the spring, which, for a particular setting, is a constant, K, and the force exerted on the valve, $P_{out} \cdot a$, where a is the effective area of the valve.   Upward pressure is the sum of the force on the diaphragm, $P_{out} \cdot A$, where A is the area of the diaphragm, and the force on the valve, $P_{in} \cdot a$.   Pressures are gauge, rather than absolute values.

$$K + P_{out} \cdot a \;=\; P_{out} \cdot A + P_{in} \cdot a$$

$$P_{out} \;=\; (K - P_{in} \cdot a)/(A - a)$$

Thus, any change in P_{in} results in a change in the opposite sense to P_{out}, by an amount which depends upon the setting and the geometry of the regulator.

Gases are usually supplied in cylinders which are filled to 17MPa (2500 psi), and can be used down to the pressures used in chromatography, i.e. 200 – 400kPa (30-60 psi).   This is a very large change, and however well designed a pressure regulator may be its outlet pressure will vary significantly as the cylinder empties.   This can be overcome by using a two-stage regulator: the first, preset, stage reduces the cylinder pressure to say 1.4MPa (200 psi), and the second stage sets some lower pressure from this.   This provides a pressure which is sufficiently constant for most purposes during the lifetime of a cylinder.

If a single cylinder supplies gas to more than one point, the cylinder regulator should be set to a value of about 100kPa (15 psi) above that required at any point, and a further pressure regulator installed wherever the gas is to be used.   This both provides an extra stage of control and ensures that variations of flow or pressure in any one application will not feed back to the others.

2.2.2   Mass Flow Controllers

If the pneumatic resistance of any downstream device varies, the pressure regulator will not provide a constant flow.   For this purpose, a mass flow controller should be used downstream of the pressure regulator.

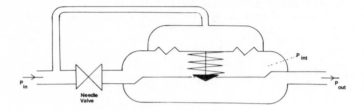

Fig. 2.2   Mass flow controller

Figure 2.2 is a schematic diagram of such a device.   The spring holds the internal valve open by pushing the diaphragm up with a force K, which would be balanced by a pressure of about 35kPa (5 psi) above the diaphragm.   When a constant inlet pressure P_{in}, which is greater than 35kPa (5 psi) is applied, the internal valve closes.   Gas enters the chamber beneath the diaphragm at a rate which depends upon the setting of the needle valve until the internal pressure, P_{int}, and the spring force balance the inlet pressure. At this point the internal valve opens, releasing gas to the outlet at a rate such as to maintain P_{int}.   So long as the downstream resistance does not increase so as to cause P_{out} to equal P_{int}, the condition for constant flow, i.e. a constant pressure drop, P_{in} – P_{int}, across a resistance (the needle valve) is satisfied.

As with the pressure regulator, there is an error term due to the pressure drop across the internal valve.   If the areas of diaphragm and valve are A and a respectively, the balance of forces is:

$$P_{in} \cdot A + P_{int} \cdot a = P_{int} \cdot A + K + P_{out} \cdot a$$

$$P_{in} = P_{int} = K_1 + (P_{out} - P_{int})\, a/a$$

This means that the pressure drop does vary with outlet pressure, but the effect can be minimised by careful design.

In an alternative configuration, the needle valve is replaced by a fixed restriction, and the flow controlled by altering the tension on the spring.

The electronic mass flow controller works on an entirely different principle.   A thermal flow sensor compares the temperature difference upstream and downstream of a heater filament in the gas.   This signal is compared with a reference voltage equivalent to the desired flow and any imbalance is used to adjust a control valve.   The device is insensitive to changes in upstream or downstream pressure, except that the inlet pressure must, of course, be high enough to overcome the downstream resistance. This type of device is necessary where chromatographic parameters are entered through a keyboard.

2.2.3   Pressure and Flow Measurement

Bourdon-tube gauges are the most common means of measuring pressure. They are very commonly fitted to pressure regulators, and are often used down-stream of flow controllers to indicate column back-pressure. Precision gauges are used for some purposes, but normal pressure gauges should be regarded as indicators rather than accurate measurement devices: they only give an accurate reading at about half full scale, the dials are too small for precise readings, and the pointers can stick, requiring a sharp tap before taking a reading.

A variable area flow meter (rotameter) consists of a glass tube of gradually tapering internal section, in which a float is supported in a position which depends upon the upward flow of gas past it. The flows commonly used in gas chromatography are at the lower limit of measurement by this means, and so they should also be regarded as indicators rather than accurate meters. The reading depends upon the flow, viscosity and pressure of the gas: they should not be used in areas where the pressure may change, and the calibration will vary if a different gas is used.

Electronic mass flow controllers incorporate a thermal flow meter, which commonly has its own digital output. These are considerably more precise than variable area flowmeters.

Soap film flowmeters, where a soap bubble is timed between two marks on a glass tube, provide very good measurement of the volumetric gas flow at ambient conditions. This type of flowmeter is used at the outlet of the column or the detector. Any change in the pressure or flow feeding the system will take a little while to settle down before the flow at the outlet steadies at the new value. This time will generally be less than a minute, even for long packed columns.

The average linear carrier gas velocity can be measured by timing the elution of an unadsorbed component. This is a more useful measure than volumetric flowrate if one is setting up a column for maximum efficiency, and can be made with similar precision to that of a soap film flowmeter. Columns used in gas analysis, of course, specialise in separation of components which are unretained on most other columns, and so one cannot say with absolute certainty that even the first peak to emerge is completely unadsorbed. Choice of helium as the unadsorbed component will minimise the error. If helium is the carrier gas hydrogen can be used; this produces a larger, but still barely significant error.

2.2.4   Choice of Method of Control and Measurement

Carrier gas supplied at constant pressure to a column at constant tempera-ture will flow through the column at a constant rate. Furthermore, small leaks between the pressure regulator and column will not affect the column flow: the regulator can provide enough flow to cater for both the leak and the column without the applied pressure being significantly altered. When a pneumatic mass flow controller is used, it must be preceded by a pressure regulator, and so the reasons for introducing this extra complication should be well thought out.

The most common use for flow controllers involves column temperature programming. As the temperature rises, so does the gas viscosity and hence the pneumatic resistance. The flow controller maintains a constant flow in spite of this varying resistance. This means that the gas velocity in the column increases during the programme, due to thermal expansion of the carrier gas. Deans (1968a) has shown that there is less variation of gas velocity when a column is temperature programmed at constant pressure rather than at constant flow. The importance of constant mass flow concerns the detector, not the column. A detector such as the katharometer is flow sensitive, and so constant mass flow is desirable for a steady baseline. The flame ionisation detector will not display baseline drift due to flow variations, but its response varies with the ratio of carrier gas to hydrogen entering the flame; constant mass flow maintains this ratio.

The pneumatic mass flow controller has a number of disadvantages. Since it contains a needle valve or fixed orifice, it is much more temperature sensitive than a pressure regulator, and should be fitted in a constant temperature enclosure. Any small leak between the controller and the column will cause the flow to be shared between the leak and the column, reducing the column flow. It is slow to respond to disturbances, such as those produced by valve operations. This latter point is particularly relevant in gas analysis, where a gas sampling valve (see 2.3) is fitted between the flow controller and the column. Injection of a sample introduces the sample loop, which contains gas usually at atmospheric pressure, into the carrier gas line, which is at a higher pressure. This causes a disturbance in the carrier gas flow which lasts until the loop volume has been pressurised to the column inlet pressure. This occurs slowly when using flow control, but rapidly when using pressure control. Figure 2.3 shows the result of using each form of control with a flow sensitive detector.

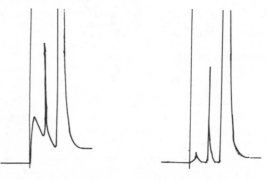

Flow control   Pressure control

Fig. 2.3   Baseline disturbance on injection

Column switching can be performed using valves (see 5.3) or pressure balancing techniques (see 5.4). When using valves, a mass flow controller can be used to overcome changes in pneumatic resistance caused by column isolation and by-pass, but the same argument applies as when using a gas

sampling valve.   The flow disturbance will be of much shorter duration if a
pressure regulator is used and a needle valve in the column by-pass line is
adjusted to have the same resistance as the column.   When using the
pressure balancing technique, pressure regulation is mandatory.

In summary, except for applications which require temperature programming,
pressure control should be preferred.   If a chromatograph is required to
perform occasional temperature programmed analyses, then the carrier gas
supply to each column should consist of a pressure regulator followed by a
mass flow controller, with an on/off valve fitted in parallel with the flow
controller.   When flow control is not needed, this by-pass allows the
column to be operated under pressure control.

A Bourdon-type gauge is used to indicate the outlet value set by a pressure
regulator, but gives no other diagnostic information.   If there is a leak
or a blockage between the regulator and the column, this may be sensed by a
variable area flowmeter situated in a constant pressure zone before the
regulator.   Variable area flowmeters themselves are often fitted before
mass flow controllers, but again only indicate the setting.   A pressure
gauge fitted after the flow controller will show an abnormally high or low
back-pressure due to obstruction or leakage.   Figure 2.4 shows the layout
of a gas flow control system incorporating all these features.

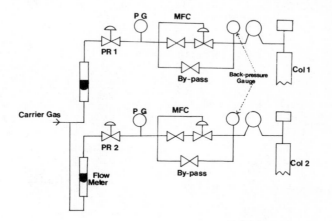

Fig. 2.4.   Gas flow control system

Auxiliary gases such as hydrogen and air for a flame ionisation detector are
usually controlled by varying the pressure applied to a fixed restrictor.
Where these flows need to be stable, as in the case of hydrogen to the flame
ionisation detector, the restrictors should be located in an area which
ideally is thermostatically controlled, but is at least free from temperature variations such as may be caused by temperature programming of the
column oven.

2.3   SAMPLE INTRODUCTION

All chromatographs are provided with injectors for syringe injection of

liquids, almost invariably of the septum type.   This type of injector can
be and is used for introduction of gas samples, using gas-tight syringes,
but the majority of samples are introduced via a gas sampling valve or
by-pass injector.   These devices, which allow excellent sample size
repeatability, come in various configurations but share a common principle;
the gas sample is introduced into a defined volume, at a known temperature
and pressure (commonly ambient), and the carrier gas then sweeps the entire
contents of this volume into the column.   The defined volume, or sample
loop may be changed, usually in the range 0.1 to 5 mls.

6-port valves are most commonly used for gas sampling.   8 or 10 port valves
can be used by plugging or joining unused ports, but are more usually chosen
when gas sampling is combined with some other function (see 5.3).

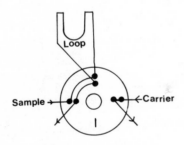

Configuration 1

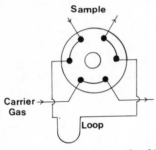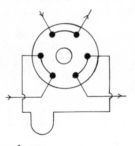

Configuration 2

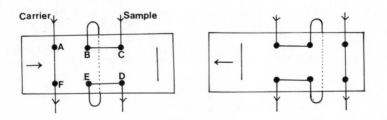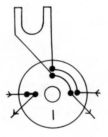

Configuration 3

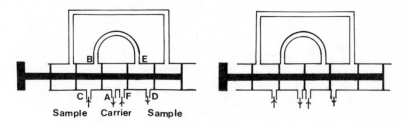

Configuration 4

Fig. 2.5.   Gas sampling valves

Figure 2.5 shows four configurations of sampling valve.   In each case the
left hand position is "load" with sample purging the loop, and the right
hand position is "inject".   Configurations 1, 2 and 3 use a metal block
containing the ports and a filled polytetrafluoroethylene rotor or slider in
which the connecting grooves are machined.   2 and 3 use the same orientation arranged for rotary and linear operation respectively.   4 uses a
piston and cylinder with O-ring seals and push-pull operation.   It also
needs an extra pair of ports and an external jumper if sample flow is not to
be interrupted.

Sampling valves are usually operated at ambient temperature, although valves
are available which will withstand temperature up to 300°C.   Any of the
types of valve described are effectively leak-tight at the pressures
normally used in gas chromatography (up to 700 kPa, 100 psig), but particular valves are available which will withstand much greater pressures.

Gas sampling valves may be mounted in the column oven, in a separately
heated adjacent oven, or externally.   They may be fitted before or after a
septum injector, or in place of it.   Ideally there should be as little dead
volume as possible between the valve and the column, so that the sample may
enter the column as a narrow band.   Sharp bends and changes in diameter
should be avoided, as these cause tailing of the sample band.   The choice
of location will usually be a compromise which depends upon other applications envisaged for the chromatograph.

Mounting the valve in the column oven allows the minimum dead volume but
limits the column temperature.   If it is in a separate oven, temperature
stability is improved over external mounting, but accessibility for changing
sample loops is poor.   If it is mounted after a septum injector, the dead
volume is reduced, but the range of liquid samples which can be analysed is
restricted by the upper temperature limitation on the valve.   The most
flexible arrangement consists of the valve mounted in a separate oven
upstream of a suitable septum injector.   The normal vaporising injector is
not the best design in this case.   They contain various changes in diameter
and a rather tortuous flow path to prevent sample blow-back.   An injector
designed for on-column injection, such as that in Fig. 2.6 is preferable.

Samples may be purged through the valve, in which case a sufficient volume
must be passed to ensure that all traces of the previous sample have been
removed: twenty times the volume of the valve and connecting tubing are
considered adequate for this purpose.   Alternatively, where the amount of

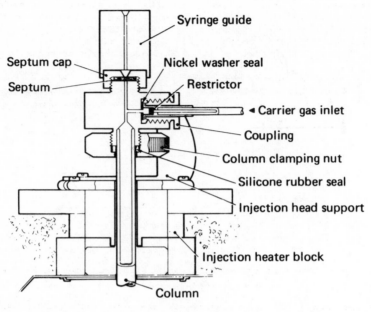

Fig. 2.6   On-column injector (Pye Unicam Ltd.)

sample is limited, the valve and connecting tubing may be evacuated, and the sample then allowed to fill the evacuated space.   Figure 2.7 shows a configuration which allows either method to be used.   The connecting lines should be of small diameter ($1/16$" o.d. tubing is suitable), and if there is any danger of particulates being carried into the valve, a filter, such as a sintered disc, should be included in the sample line.

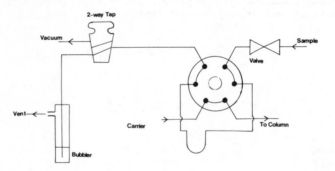

Fig. 2.7.   Transfer system for samples at above atmospheric pressure

Sample size repeatability depends upon the temperature and pressure of sample in the loop.   These can be measured, but it is much easier to ensure

that the standards and samples reach the same temperature and pressure
before injection.   The temperature may be that of an oven or it may be
ambient.   The pressure will almost always be atmospheric.   Over the short
term (2-3 hours) these are not likely to vary sufficiently to matter, but
for very accurate work their possible influence should be borne in mind.

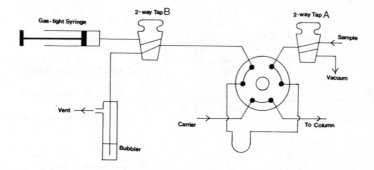

Fig. 2.8.   Transfer system for samples at below atmospheric pressure

If the gas sample is only available at pressure less than one atmosphere,
the evacuation technique can be used, the available sample then being shared
between the container and the sample loop, and its pressure measured.   This
means that the uncertainty of pressure measurement must be added to the
uncertainty of the chromatographic determination.   Figure 2.8 shows a
modification of the layout which will allow the sample to be injected at one
atmosphere.   The valve is evacuated through tap A, which may be a double
oblique bore stopcock on a gas sampling tube, or a similar tap inserted in
the line, up to and including the gas tight syringe.   Tap A is then rotated
so that the sample container is connected to the valve, and the plunger of
the syringe withdrawn so that more sample is drawn from the container.   Tap
A is now closed, and the syringe plunger pushed in to pressurise the sample.
Tap B is rotated, and excess pressure vented.

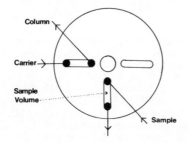

Fig. 2.9.   Sample valve for liquefied gases

Gases which are liquefied under pressure at ambient temperature may be fully
vaporised or alternatively handled as a liquid.   In the latter case, the

volume of sample is very much reduced (typically 200-400 times), and sample sizes in the microlitre range are required.   Fortunately, there is no shortage of appropriate valves due to the popularity of liquid chromatography and Fig. 2.9 shows an example.   In this case the valve must be kept at ambient temperature - any increase in temperature causes a large increase in vapour pressure, making it difficult to maintain a liquid phase sample in the valve.   Above the critical temperature of the sample it is, of course, impossible.

2.4   COLUMNS AND COLUMN OVENS

Column packing materials are the subject of Chapter 3, and the applications to which they are put are described in Chapter 6.   Column tubing is most commonly stainless steel.   Aluminium is preferred for sulphur compounds, with PTFE being needed for extremely low levels.   PTFE is also used for samples containing corrosive gases, and PTFE or glass are suitable for water vapour measurement.

The column oven requirements are substantially the same as those common to other applications, but two points are worth emphasis.   The first is that multiple columns and valves are more commonly used in gas analysis, and so ample space and good accessibility are desirable.   The second point relates to the fact that many gas analyses require temperatures close to ambient: heat input from injector and detector zones, and from the fan motor can prevent some ovens from controlling in this region.

The circulating air oven is the more common type, and offers considerable advantages in terms of the speed with which a new temperature can be set. It is, of course, essential for temperature programmed or subambient operation, and more recent models incorporate means of venting air from the oven and replacing it with air at ambient temperature as part of the temperature control mechanism.   This allows stable operation at $10-20^{\circ}C$ above ambient.   If only isothermal operation is required, then use of a conductively heated oven should be considered.   This type of oven uses a massive metal plate, in which the heaters and temperature sensors are embedded, and on which the components parts are mounted.   Response to upsets, such as selecting a new isothermal temperature or changing a column, is slow, but the stability of the selected temperature is excellent.

2.5   DETECTORS

Choice of an appropriate detector is important in gas analysis, but it is not the only variable parameter.   If a trace component is obscured by a neighbouring major peak then use of a more sensitive universal detector will not help.   A selective detector will only be useful if it selects the trace rather than the major component.   The more selective a detector, the more one should query the need for the added complication of a chromatographic separation so as to present components to the detector serially rather than in parallel.   Indeed, some detectors can be used in this way, continuously sampling an atmosphere, but connected in a way which allows chromatography to be invoked as an aid to identification.   Occasionally, use of an appropriate detector may be the only change necessary to solve a particular analytical problem, but more usually it must be combined with choice of an appropriate column or columns (see Chapter 3) and appropriate separation tactics (see Chapter 5).

The thermal conductivity detector (TCD) and flame ionisation detector (FID) are the most commonly used in gas analysis.   The electron capture detector (ECD) is becoming more popular particularly in the sensitised mode. Ultrasonic, flame photometric and helium ionisation detectors are used for particular applications.   The photoionisation detector, although responding to a wide variety of compounds, has so far been little used for gases.

David (1974) has reviewed chromatographic detectors in considerable detail. Novak (1975) has described their quantitative behaviour, using the classification into mass- and concentration-sensitive detectors devised by Halasz (1964).

2.5.1   Thermal Conductivity Detector (TCD)

The detector consists of an electrically heated element in a thermostatted metal block.   Column effluent passes over the element, which in many cases may be arranged in pairs.   The detection principle is the change in the thermal conductivity of the gas flowing through the detector when a component elutes from the column, and the mechanism relies upon measurement of the rate of heat loss from the element.   Heat is lost due to conduction through and convection by the column effluent, by radiation, and by conduction through the supporting wires.   The largest of these components is conduction through the gas, and it is changes in thermal conductivity with composition which most affect the detector output.

Because of the extreme temperature sensitivity of the device, the traditional method of construction of a TCD involves a duplicate element or pair of elements mounted in the same block, over which pure carrier gas flows.   This gives a measure of compensation for changes in block temperature, and in the temperature, pressure or flow of carrier gas.   The different elements are referred to as the "sensing" and "reference" elements.   Their roles are, of course, interchangeable, the only requirement being that the element nominated as "reference" should only have pure carrier gas flowing over it.

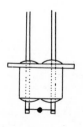 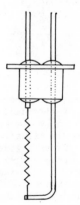

Fig. 2.10.   Detector elements, (a) thermistor, (b) filament

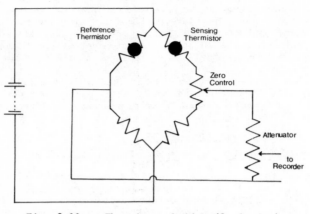

Fig. 2.11.   Thermistor bridge (2-element)

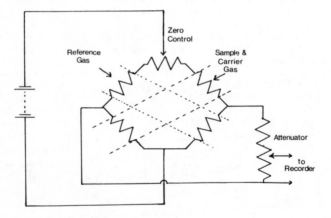

Fig. 2.12.   Filament bridge (4-element)

The elements are either wire filaments - platinum, tungsten or tungsten alloy are common - or thermistors.   Filaments are usually helically wound, 1 cm in length and 0.5 mm in diameter, and thermistors are typically 0.5 mm beads.   Figure 2.10 illustrates a thermistor and a filament detector element.   The thermistor allows higher sensitivity at lower detector block temperatures (less than 100°C typically), and the possibility of a low detector volume, which for the same carrier flow rate means a low time constant.   The detector elements are arranged in a Wheatstone bridge, which senses the imbalance when a component peak passes through.   Figure 2.11 shows a bridge in which two of the arms are the sensing and reference thermistors, and Fig. 2.12 shows a four-element bridge.   Figure 2.13 illustrates the layout of a four-element detector-block.   The column effluent entering the cell is split so that approximately 25% passes over

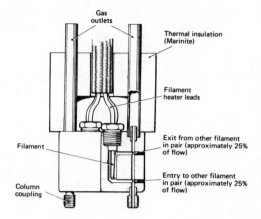

Fig. 2.13.   Four-element detector block (Pye Unicam Ltd.)

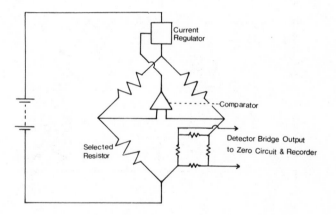

Fig. 2.14.   Constant-temperature bridge (Perkin-Elmer Ltd.)

each element, and 50% by-passes them.   Since the detector is concentration-
sensitive the sensitivity is not affected, and the design reduces the
response to flow rate changes.   Fairly high flowrates are necessary, however, to avoid a large time constant.

If a bridge is controlled so as to maintain a constant temperature of the
sensing elements rather than a constant bridge voltage, there is an advantage in sensitivity, in linear range and in time constant (Wittebrood 1972).
Figure 2.14 is a schematic of such a system.   The imbalance signal is
amplified and fed back to the bridge so as to control the filament temperature.   It is more complex than the original bridge, but its advantages have
caused it to be widely adopted by the major suppliers.

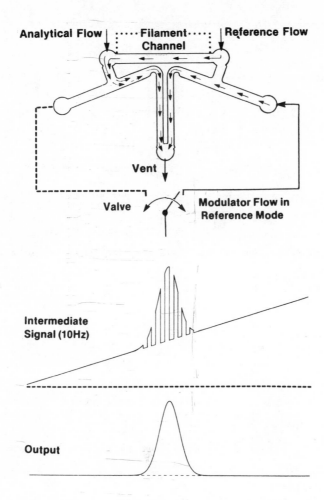

Fig. 2.15.   Operation of modulated single-filament detector
              (Hewlett-Packard Ltd).

A recent alternative design of thermal conductivity detector (Craven and Clouser, 1979) uses a single filament with a pneumatic switching device which exposes it alternately to the column effluent and reference flows at a frequency of 10 cycles per second.   This is illustrated schematically in Fig.   2.15.   The signal is sensed by demodulating the output at the switching frequency.   This has the advantage that the signal can be processed digitally throughout.   A further benefit is more stable baselines. With a conventional detector, small changes in detector block temperature cause drift, whereas with this new design, since the drift is not modulated at 10 cycles per second, it will not show.   This may be a two edged sword, however.   A steady baseline from a conventional detector implies stable conditions and one would expect good repeatability; if drift cannot be seen because of the design, it cannot serve as a diagnostic function.

The size of the signal from a thermal conductivity detector is generally in the range 10 µV to 1V, and is amenable to direct recording and integration. The detector responds to any component with a thermal conductivity different to that of the carrier gas, which has more significance when the samples are gases than for liquids or solids. The role of the carrier gas is described in Chapter 4. The concentration of any individual component which can be detected is generally in the range 50 ppm to 100% molar.

2.5.2   Flame Ionisation Detector (FID)

The most widely used detector in gas chromatography generally, it can be regarded as a selective detector for certain types of gas analysis. The detection principle is that of chemi-ionisation of organic compounds on combustion. The mechanism involves mixing the column effluent with hydrogen, burning at a jet, and collection and measurement of virtually all the ions formed. Figure 2.16 shows a modern design, in which the cylindrical shape of the collector electrode and the applied potential (-150V) ensure efficient collection. It should be remembered that the effective volume of the detector is the volume of the flame itself, not that of the space between jet and collector. As soon as an ion is formed in the flame, its fate is sealed.

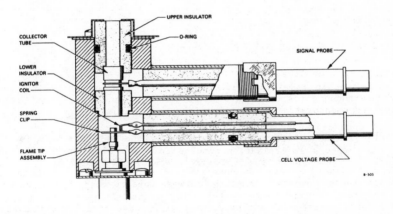

Fig. 2.16.   Flame ionisation detector (Varian Associates Ltd.)

The fact that there is very little difference between designs from various manufacturers is a tribute to the elegance of the original concept. Early work such as that of Sternberg, Gallaway and Jones (1962) on relative response to different compounds and classes of compounds is still generally valid. This shows that hydrocarbons have molar responses which are proportional to carbon number and that heteroatoms reduce this response in a broadly consistent manner. This means that it is possible to calculate the approximate response of a compound by summing the effective carbon number

contributions of the atoms and groupings which comprise it. Relative
responses show some variation between detectors, and can vary with operating
conditions on the same detector. For accurate work, they should be checked
for each instrument and for each set of conditions.

The flame ionisation detector responds to virtually all organic compounds.
It does not respond to the rare gases, O_2, N_2, CO, CO_2, H_2, H_2S,
COS, halogens, hydrogen halides, ammonia or oxides of nitrogen. The oxides
of carbon can be detected if they are converted catalytically to methane
after separation on the column (see 2.6.1).

Detector sensitivity depends on the auxiliary gas flows. At a particular
carrier gas flow, there is an optimum hydrogen flow which maximises response
This varies between detectors but tends to occur when the volumetric
flowrates are similar. Response rises to a plateau with increasing air
flow. This is achieved when the rate of air diffusion into the flame is no
longer limited by the rate of supply into the detector.

The size of current passed by the detector is extremely small (down to
10^{-13} amps) and can only be measured by using an amplifier as an impedance matching device. The usable range of the detector covers 6 or 7
orders of magnitude, which is greater than the range of the amplifier. To
deal with this, it is common to change the input range of the amplifier by
factors of 10, 100 and 1000, so that within each range, a signal of between
10 V and 1V can be supplied to an integrator, or via an attenuator to a
recorder. Some recent chromatographs have an auto-ranging amplifier, which
takes full advantage of the operating range of the detector.

The detector is destructive and mass-sensitive – its response depends on the
rate at which a sample component enters it. For hydrocarbons, the limit of
detection is of the order of 10^{-11}g of carbon/sec. This means that
the concentration range covered in a gas sample can be 0.1 ppm to 100%
molar.

2.5.3   Electron Capture Detector (ECD)

Hitherto, the electron capture detector has been little used in gas analysis
but recent developments could make it much more popular. The detection
principle is that of electron attachment by the solute molecules. The
mechanism involves an ionisation chamber in which low-energy electrons are
released from the carrier gas, creating a standing current.

$$\beta + N_2 \longrightarrow N_2^+ + \varepsilon$$

Certain solutes will combine with the electrons to form stable negative
ions.

$$AB + \varepsilon \longleftrightarrow AB^-$$

The negative ions can undergo further reactions, including dissociation, but
the most common is collision with a positively charged carrier gas ion.

$$AB^- + N_2^+ \longrightarrow \text{neutral products}$$

This collision is more likely to occur than a recombination between an
electron and a carrier gas ion, and so the result is a drop in standing
current. If the collision merely discharged the ions so as to leave the

original solute and carrier gas molecules, then the process could start again, with each solute molecule removing several electrons from the system. Those materials which capture electrons to form negative ions are invariably polyatomic, and the recombination reaction is sufficiently energetic to fragment them. The fragments themselves could be electron-capturing, but this is unlikely as relatively few molecules have this property.

Oxygen is mildly electron-capturing and sulphur hexafluoride intensely so. The ratio of their responses is of the order of 10^6. Other than these, relatively few gases respond in the detector. Examples are oxides of nitrogen and halocarbons such as CF_2Cl_2 and $CBrClF_2$.

Figure 2.17 shows a typical design. The detector cell contains a β-emitter, either Tritum (adsorbed into titanium foil), or, more commonly in modern instruments, Nickel-63. The latter is preferred on safety grounds and because it allows the detector to be operated at higher temperature (up to 350^oC), which both enhances response and reduces contamination due to column bleed. Voltage is applied to the cell as a series of pulses. Charged particles build up in the detector between pulses, and are collected during them. Since oxygen, while only mildly electron-capturing, is a major component of air, the gas outlet line must prevent back-diffusion.

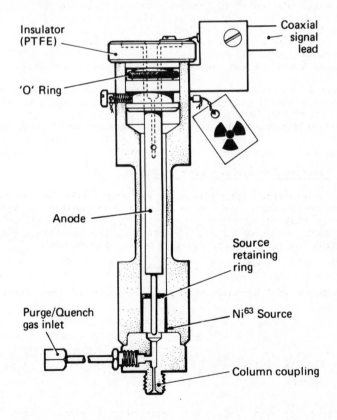

Fig. 2.17. Electron capture detector (Pye-Unicam Ltd)

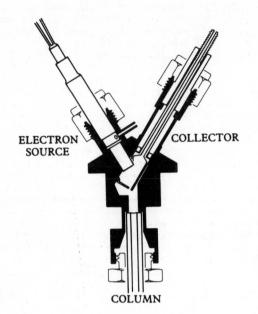

ELECTRON   COLLECTOR
SOURCE

COLUMN

Fig. 2.18.   Non-radioactive ECD (Hewlett-Packard Ltd.)

A non-radioactive detector has been described by Sullivan (1979), and is
illustrated in Fig. 2.18.   The source of electrons is a filament.   Most
electrons are collected on the grid, but some drift through with the purge
gas supply into the detector cell.   This design is attractive not only
because it does not contain a radioactive source, but also because it could
be used at higher temperatures, further enhancing response.   The mechanism
of this detector has not been thoroughly described, and so the remaining
comments in this section relate to radioactive designs.

Carrier gas is usually nitrogen or argon/methane, 95/5%.   The methane acts
as a moderator, ensuring that the electrons are in thermal equilibrium with
the carrier gas.   Argon/methane is less commonly used, although it gives
rather greater response, which may be in part due to the inconvenience of
having a specialised carrier gas for this application alone.   Argon/ methane
is completely incompatible, for example, with simultaneous use of a flame
ionisation detector.

Lovelock, Maggs and Adlard (1971) describe coulometric response for SF_6
and various fully halogenated hydrocarbons.   It is favoured by low carrier
gas flowrates and long pulse intervals - both of which allow a longer time
for electron attachment to take place.   When complete ionisation occurs,
components will have the same response, and the detector responds to the
rate of mass input.   At low ionisation efficiencies, which are associated
with high flowrates, the detector acts as a concentration - sensitive device.   A corollary of coulometric response is that the component so detected is quantitatively destroyed in the detector.   By switching a D.C. voltage on and off to such a detector, highly electrophilic components can be
allowed to pass through the detector or not.   Lovelock (1975) fitted two
ECDs in series at the column exit, and used the first as a switch, the

second as the chromatographic detector.   By operating the switch at a fixed frequency, and demodulating the signal from the second ECD at the same frequency, a considerable increase in sensitivity was shown.   The effect increases the discrimination between strongly and weakly electron-capturing species and also eliminates non-modulated signals, such as baseline drift.

When operating in the concentration-sensitive mode, the detector is inherently non-linear, since increasing concentrations of solute encounter decreasing concentrations of electrons with which to react.   Maggs and co-workers (1971) compared constant pulse frequency with variable pulse frequency operation.   In the latter mode, a constant current is maintained in the cell by increasing the pulse frequency with concentration of component.   The change in pulse frequency is the measured quantity, being a linear function of concentration over a range of 5×10^4.   This method of operating ECDs is now almost universally used by instrument manufacturers.

The most significant recent publications on the ECD refer to sensitisation of the detector to gases to which it is not normally responsive.   Simmonds (1978) used an ECD for ambient levels of N_2O, and showed that addition of 100 ppm oxygen to the nitrogen carrier gas allowed simultaneous determination of ambient levels of CO_2 without adverse affect on the N_2O sensitivity.   The mechanism of response is different, as is evident from the effect of detector temperature on sensitivity. N_2O response increases with temperature, whereas that for CO_2 peaks at 220°C, and falls off to zero by 350°C.   Since oxygen is itself electron capturing, its continuous presence in the detector reduces the standing current (in the constant pulse-frequency mode), but only by 10%, which is acceptable. Miller and Grimsrud (1979) used a constant-current detector, and levels of oxygen in the carrier of up to 2000 ppm.   The enhancement of peaks for vinyl chloride and other halocarbons was considerable (up to 200-fold) and an approximately linear function of oxygen concentration.

Phillips and co-workers (1979) used varying amounts of N_2O in nitrogen carrier gas, for analysis of low concentrations of H_2, CH_4 and CO_2. Again the response increased with the concentration of N_2O, up to about 10 ppm.   The chemistry in the detector is governed by the reactions.

$$\varepsilon \quad + N_2O \longrightarrow O^- + N_2$$

$$O^- + N_2O \longrightarrow NO^- + NO$$

$$NO^- + N_2 \longrightarrow NO + N_2 + \varepsilon$$

The steady-state concentrations of electrons, O^- and NO^- ions depend upon the rate constants of these three reactions.   Solutes entering the detector can thus react with electrons, O^- or NO^- ions, to form stable negative ions, which are   removed in the normal way by collision with positively charged carrier gas ions. H_2, CH_4 and CO_2, which do not attach electrons, all react with O^- and may be detected with high sensitivity.   Sievers and co-workers (1979) used 20 ppm N_2O in nitrogen carrier gas, and showed picogram sensitivity for hydrocarbons.   The response to normally electron-capturing halocarbons is reduced when N_2O doped carrier gas is used, but this reduction is much less than the enhancement of response for components which are not normally electron-capturing.   This would allow rapid if tentative identification of compound types, by switching carrier gas between pure N_2 and N_2O-doped nitrogen.

Sensitisation of the ECD by doping the carrier gas considerably expands the range of application of the detector in gas analysis.   Continuous development is expected in this most interesting area.

2.5.4   Flame Photometric Detector (FPD)

This detector allows sensitive and selective measurement of compounds containing sulphur or phosphorus.   The detection principle is the formation of excited S_2* or HPO* species in a reducing flame.   The mechanism involves measurement by a photomultiplier tube of the characteristic chemiluminescent emission from these species.

The first use as a chromatographic detector was described by Brody and Chaney (1966).   The column effluent is mixed with oxygen and burned in an excess of hydrogen.   The optical filter can be changed, to allow the photomultiplier to view light of 394 nm for sulphur mesaurement or 526 nm for phosphorus.   A more recent dual version allows simultaneous phosphorus and sulphur measurement with two photomultipliers viewing the flame through different filters.

A drawback to this design is the tendency for the flame to extinguish by flashing back down the inlet line when a large amount of component elutes from the column.   Gibbons and Goode (1968) described a simple non-extinguishing version for gas analysis, in which the column effluent was mixed with the hydrogen.   Figure 2.19 illustrates the detector.   Most chromatograph manufacturers can now supply an FPD and use non-extinguishing designs.   Air has replaced oxygen in many of the new detectors.   This is convenient since air is more usually available in a chromatographic laboratory but does not give such good sensitivity as from a hydrogen/oxygen flame.

There are very few phosphorus-containing gases, and few references to use of the FPD to determine them.   Sulphur containing gases, on the other hand, account for a substantial number of applications of the FPD, and for a scarcely smaller number of opinions about the mechanism of response.   It is accepted that the response is a complex function, of type

$$y = a.x^b$$

where y is the instantaneous system response
x is the rate of entry of a sulphur compound into the detector
and a, b are constants.

b is stated to vary between 1.5 and 2, depending upon the compound, the detector and the conditions.   It is also stated that at the same rate of entry of different sulphur compounds, the response per unit of sulphur differs widely.   In fact these arguments are inseparable, as if the exponent, b, varies between two compounds, there is a maximum of one point at which the response per unit of sulphur could be the same.   In the authors' view, many if not all of these interpretations stem from the nature of the FPD response, and are not rigorously justified.

Chemiluminescent response occurs above or in the top area of the flame, which means that species present in this zone have passed entirely through the flame.   It seems unlikely that any trace of the original orientation of the sulphur atom in the molecule would survive.   There is no evidence that sulphur combined in such different ways as in H_2S and SO_2 behaves

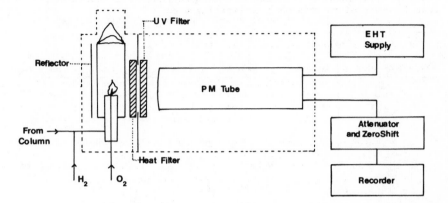

Fig. 2.19.   Flame photometric detector

differently, and it has been shown that adjacent sulphur atoms, as in disulphides, do not stay attached so as to form S_2^*. The concentration of S_2^* in the flame must be proportional to the square of the concentration of S. If the photomultiplier output is linearly related to the concentration of S_2^* then the response must be proportional to the square of the concentration of sulphur entering the flame. Any value of b other than 2 should be investigated.

Many sulphur-containing gases are quite strongly adsorptive, and can suffer severe tailing or even total loss in chromatographic columns and connecting tubing. Thus it is possible to have variable levels of sulphur background entering the detector due to tailing. Even when such a background signal is electrically compensated, it has a complex effect, since (background + signal)^2 is not the same as (background)^2 + (signal)^2. Table 2.1 shows the effect of different background sulphur levels on the response to peaks of different sizes. The peak heights are those measured above the background level, and the figures in bold type are the values of the exponent, b, between the two adjacent peaks. Two effects are evident - the response increases significantly with the background, and the observed exponent varies between 1 and 2. More significantly, the background need only be small relative to the peaks for the exponent to vary over the range 1.7 to 2.0 which has caused many authors to speculate about the effects of molecular structure.

A further complication is caused by quenching - the reduction in sulphur response when a non-sulphur-containing organic component is co-eluting. The effect is also complex and unpredictable, varying both with the level of sulphur and the level of quenching agent. The sulphur response can be completely suppressed to such an extent so as to include any sulphur background, producing a negative peak. In the absence of background, very high levels of, for example, hydrocarbons can themselves produce a positive response.

Table 2.1   Effect of background on FPD Response

Background S level (arbitrary units)	Sulphur level @ peak maximum (arbitrary units)								
	1 Pk.Ht	b	3 Pk.Ht	b	10 Pk.Ht	b	30 Pk.Ht	b	100 Pk.Ht
0	1	2.0	9	2.0	100	2.0	900	2.0	10000
1	3	1.46	15	1.74	120	1.89	960	1.96	10200
3	7	1.23	27	1.48	160	1.74	1080	1.90	10600
10	21	1.08	69	1.22	300	1.46	1500	1.73	12000
30	61	1.03	189	1.09	700	1.23	2700	1.48	16000
100	201	1.01	609	1.03	2100	1.08	6900	1.22	30000

A dual flame detector was described by Rupprecht and Phillips (1969), designed to overcome quenching.   The lower flame was oxygen-rich, and the upper flame reducing. Components entering the detectors were converted to CO_2, H_2O and SO_2, and the S_2^* emission in the upper flame was free from quenching effects.   This was successful, but suffered operational problems.   More recently, a commercially available version has been described (Patterson, Howe and Abu-Shumays, 1978, Patterson, 1978) in which both flames are reducing.   Organic molecules are broken down in the lower flame into simpler species, which, without being defined, do not cause quenching when they enter the upper flame.   Chemiluminescence occurs in both flames, but only the upper is viewed.   It can be used as a single flame detector by turning off the lower flame.   The dual flame detector is very effective at eliminating quenching, but has a somewhat lower sensitivity than a single flame detector.   Its principal disadvantage is that other species, such as hydrocarbons, have a much greater response than in a single flame detector.   Thus a peak may be a trace of sulphur compound or a larger quantity of hydrocarbon.   This can be resolved by injecting different sample sizes - hydrocarbon and sulphur responses vary with the first and second powers of sample size respectively.

In gas analysis, sulphur compounds can usually be separated without significant interference from other species, and so the extra sensitivity of a single flame detector may well be preferred over the non-quenching ability of the dual flame design.   In spite of the complications outlined above, the FPD is very popular for trace sulphur gases and can be used with a high degree of confidence once its mechanisms are understood.

2.5.5   Ultrasonic Detector

The ultrasonic detector is a universal concentration-sensitive device.   Its performance is similar to that of the thermal conductivity detector, but with considerably greater sensitivity.   The detection principle is the change in the velocity of sound with the composition of a gas stream, and the mechanism involves two transducers mounted in a cell through which the column effluent flows: one is driven by an oscillator at a fixed ultrasonic

frequency, and the other receives the signal and passes it to a network
which compares the phase angles of the two signals.   Any change in the
velocity of sound causes the phase difference to shift.   It is possible to
use two detectors as sample and reference (as with the TCD), in which case
the two received signals are compared.   Figure 2.20 indicates the design.

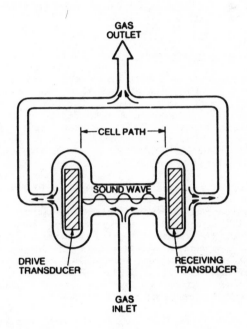

Fig. 2.20.   Ultrasonic detector (Tracor Inc.)

The detector is sensitive to all components whose molecular weights and/or
specific heats are different from those of the carrier.   Good temperature
control is vital because of its effect upon the velocity of sound, but once
this has been achieved, components can be detected with parts per million
sensitivity.   Approximate relative responses can be predicted by ignoring
specific heats and only considering differences in molecular weight, but for
accurate work each component should be calibrated.   In the original design
(Noble, Abel and Cook, 1964), the output signal returned to zero each time
the phase shift exceeded 360° or multiples thereof.   A more recent design
(Grice and David, 1967) allows up to 64 complete phase changes, and hence a
much greater linear range.   The detector cell must be pressurised to
support signal propogation, the necessary pressure varying with the carrier
gas.   This is a disadvantage, since it effectively increases the detector
cell dead volume.

When looking for trace impurities the major component of the sample can be
used as carrier gas, as with the TCD.   There is an upper limit on molecular

AGC-B*

weight of the carrier gas - SF_6, for example, completely absorbs the signal. On the other hand, oxygen can be used as carrier with no danger of damage, as may occur with a TCD. In the same way, its ability to cope with corrosive samples depends upon the materials of construction, but there are no high temperature components, such as the filaments in a TCD.

It is probable that the high price of the detector and the fact that there is only one manufacturer have combined to make it less popular than its performance seems to justify.

2.5.6   Photoionisation Detector

Irradiation of the detector cell with high-energy photons can ionise some or all of the contents. This principle has been used in the construction of several detectors (Yamane, 1964, Price and co-workers, 1968 and Rosiek, Gudowski and Lasa, 1976), in which a glow discharge in the carrier gas, usually helium, allows sensitive detection of all other permanent gases and organic compounds. More recently interest has been focussed on the detector described by Driscoll (1977), in which the contents of a small volume cell are irradiated by an ultra-violet lamp.

As originally described, the lamp emits energy of 10.2 electron volts (eV). This allows great sensitivity (up to 100 x that of an FID), with some discrimination against gases with a high ionisation potential. More recently lamps with alternative energy ratings have become available, allowing variation in the range of components detected, and in the relative response between components. Compounds of higher molecular weight invariably have lower ionisation potentials, and consequently will be detected. The ultra-violet lamps produce a range of energies, not just a sharp band, and hence components whose ionisation potential is 0.3-0.4 eV higher than the nominal lamp value can be detected.

Table 2.2 indicates the extent to which the detector can be made to be selective. With the 10.2 eV lamp, many components including inorganic gases such as arsine and phosphine, can be detected at concentrations in the region of 10 parts per billion (1 in 10^8). The other lamps do not produce the same energy intensity, but still give sensitivity greater than that of the FID.

The detector is attractive for gas analysis, particularly for identification of trace components and environmental work. Not many applications have been published to date, but its popularity is likely to grow.

2.5.7   Helium Detector

This works on the same principle as the argon detector,but the greater metastable energy of helium (19.6 as against 11.5 eV) means that it can measure all the permanent gases which were outside the scope of the argon detector (Berry, 1962). It can detect components at the part per billion (1 in 10^9) level, but it has a rather restricted linear range. Its sensitivity means that the response is a function of carrier gas purity as well as of applied voltage and to a lesser extent cell design. The detector is mass-sensitive and non-destructive, which makes it unique among those so far described.

Table 2.2   Ionisation Potentials of Gases

Component	Ionisation Potential	Lamp Energy
Nitrogen	15.6eV	
Carbon monoxide	14.0	
Carbon dioxide	12.8	
Nitrous oxide	12.8	
Methane	12.7	
Water	12.6	
Sulphur dioxide	12.3	
Oxygen	12.1	11.7
Carbonyl sulphide	12.0	
Chlorine	11.5	
Ethane	11.5	
Ethyne	11.4	
Propane	11.1	10.9
n-Butane	10.6	
Ethene	10.5	10.2
Hydrogen sulphide	10.5	
Propyne	10.4	
Ammonia	10.3	10.0
Vinyl chloride	10.0	
Propene	9.8	
Nitrogen dioxide	9.8	9.5
Methanethiol	9.4	
Nitric oxide	9.3	
Butene-1	9.2	
Dimethyl sulphide	8.9	

Lukac and Sevcik, in two papers (1972 a, b)) discuss the mechanism of the detector. Bros, Lasa and Kilarska (1974) describe a system in which the specially purified helium exits from the detector cell into another chamber which entirely surrounds the detector, and thence through a chamber surrounding the gas sampling valve. Even with these precautions to prevent air ingress, it takes several days for the system to be purged of impurities, during which time the peaks change from negative via W-shaped to positive.

Recent workers (Bros and Lasa, 1979, Andrawes and Gibson, 1978) have observed the effect of controlled levels of gaseous impurities in the carrier gas. Some tens of ppm of impurities make the response more consistent (all peaks positive) and the response curve more linear for trace gases. Andrawes, Brazell and Gibson (1980) have shown the advantage of working in the saturation region rather than the multiplication region of response, and of using high purity helium with its built-in low level impurities rather than the ultra high purity grade. This has allowed an extension of the range of application, which was usually restricted to gas analysis on adsorption columns.

Andrawes, Byers and Gibson (1981) have quantified the effect of adding different levels of hydrogen as impurity to ultra-pure helium. At 90 ppm hydrogen, the response is most consistent for other permanent gases, although the detection limit for methane, for example, has been increased from 0.4 to 4 ppb.

It is likely that the reluctance to use the helium ionisation detector stems from the need to use ultra-pure helium, and the sometimes anomalous response. The recent developments should make it a more popular detector.

2.5.8   Other Detectors

Other detectors have been used for gas analysis, usually for quite specific applications. The list is not exhaustive.

A radiofrequency discharge detector (Glover, 1976) allows sub-ppm detection of permanent gases.

Electrochemical detectors have been used for measurement of various acidic or basic compounds, or for oxidisable substances such as sulphur-containing odorants in natural gas at the ppm level (Charron and Maman, 1980). Blurton and Stetter (1978) used a PTFE bonded diffusion electrode in a cell for detection of H_2S, SO_2, NO, NO_2 and CO at sub-ppm levels.

A thermochemical pyroelectric detector (Gaglya, 1979) has been used for measurement of H_2, CO and C_2H_4 in air at concentrations of the order of 10 ppm. The detector measures the heat of combustion over a catalytic element, and has been used in the vacancy mode, with sample flowing continuously through the column and aliquots of air being injected.

2.6   AUXILIARY DEVICES

If there is an incompatability between the amount of a component at the column exit and the sensitivity or specificity of the detector, then under some circumstances the component can be transformed into a more appropriate form or environment for the detector to use. Such devices, examples of which follow, must be of low dead volume so as not to spoil peak profiles.

2.6.1   Methanator

Low levels of CO and CO_2 are not easily detected by the thermal conductivity detector. If they are converted to methane, a flame ionisation detector can be used which is both more sensitive and free from interference due to adjacent permanent gases. Methanation is conveniently carried out by mixing hydrogen with the column effluent and passing it over a nickel catalyst at 350^oC. The nickel is coated onto a support and packed into a short length of tube of similar diameter to the chromatographic column. Parts per million levels of CO and CO_2 are easily detected.

Figure 2.21 shows a typical layout. The catalyst tube is fitted into an unused injection port position, which heats it to the appropriate temperature. The normal FID hydrogen supply is split so that part goes through the methanator. Components also present in the mixture which are normally detected by FID are unaffected, except that unsaturated hydrocarbons (ethene, propene) are converted to the alkanes. This has little or no effect on the detector response, but can in any case be allowed for by the calibration procedure.

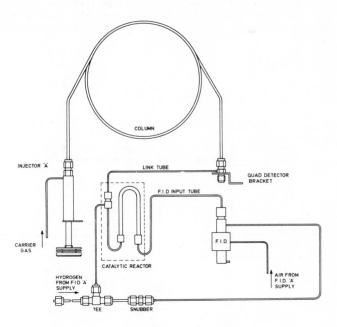

Fig. 2.21.   Methanator (Perkin-Elmer Ltd.)

2.6.2   Hydrogen Transfer System

When analysing mixtures containing permanent gases with a TCD, helium is a
popular carrier gas.   Helium does not, however, allow accurate measurement
of hydrogen content (see Chapter 4). This requires the use of a carrier gas
of low thermal conductivity, such as nitrogen or argon.   This change of
carrier gas is usually unacceptable because of the reduced sensitivity to
all other components.   The hydrogen transfer system (Johns and Berry, 1975)
uses the very high and selective permeability of palladium alloy to allow
transfer of the hydrogen as it emerges from the column into a stream of
nitrogen for detection by a second TCD.

Figure 2.22 is a schematic of the device, and Fig. 2.23 shows how it might
be incorporated into a chromatographic system.   With the palladium alloy
heated to between 500 and 625°C, the counter flow of the two carrier gases
ensures rapid and quantitative transfer.   Separation between hydrogen and
oxygen must be good, as any overlap of these components entering the
hydrogen transfer system will cause water formation, which will interfere
with the oxygen peak measurements.

2.6.3   Palladium Transmodulator

Concentration-sensitive detectors are limited in sensitivity by the fact
that successful separation of a sample of particular size requires a
particular flowrate of carrier gas which limits the maximum concentration of

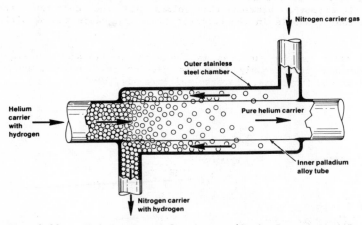

Fig. 2.22.   Hydrogen transfer system (Carle Instruments Inc.)

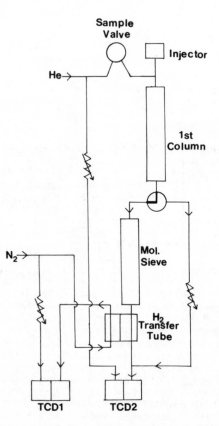

Fig. 2.23.   Hydrogen transfer system incorporated into a chromatograph

sample in carrier in the detector.   If a substantial part of the carrier
gas can be removed between the column and the detector, then the
concentration of sample in the detector rises and hence the sensitivity is
increased.

Palladium alloy can be used, as in the hydrogen transfer system, but to
remove hydrogen continuously from the carrier gas rather than from the
separated sample (Lovelock, Charlton and Simmonds, 1969).

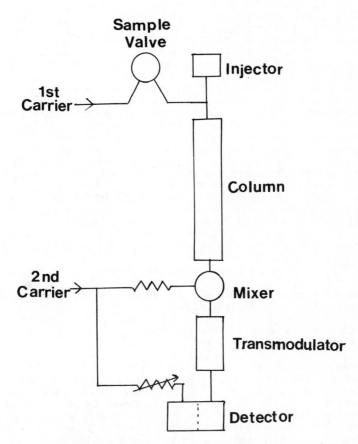

Fig. 2.24.   Palladium transmodulator incorporated into a chromatograph

The transmodulator consists of three parallel lengths of palladium silver
alloy capillary tubing, resistively heated.   Hydrogen, having diffused
through the alloy, immediately reacts with atmospheric oxygen on the outer
surface of the tubing.   This scavenging action assists the high efficiency
of the transmodulator.   The second carrier gas, helium or nitrogen, may be
blended with the hydrogen and the mixture used as carrier, or introduced
after the column with pure hydrogen used as first carrier.   The latter
method has the advantage that any flow changes in the column are not

reflected by baseline shifts from the detector.

The increase in sensitivity is related directly to the ratio of the two carrier gases. This is most conveniently increased by reducing the flow of second carrier, but the range over which this can be done is restricted by the volumes of the transmodulator and the detector. Catalytic hydrogenation of some components in the transmodulator is always likely but as with the methanator, so long as the conversion is repeatable, appropriate calibration will allow for it.

2.7   RECORDER/INTEGRATOR

There seems little point in treating these instruments separately since the price of the devices is converging and a number of modern integrators incorporate a printer/plotter, making the independent recorder unnecessary. The recorded chromatogram contains, to the experienced eye, a wealth of information which cannot be as well defined by the print-out from the most sophisticated computing integrator. This information concerns the efficiency of the separation and the performance of the chromatograph, and gives useful clues as to how a separation may be improved.   When it comes to quantitative measurement of peaks, however, the integrator, with its typical dynamic range of 10^6:1 is much superior to the recorder, with a dynamic range of 200:1.   (The range of a recorder can be increased by the use of an attenuator on the chromatograph, or of a variable range control on the recorder itself, but this assumes that the instrument is constantly attended, which is unnecessary with the integrator).   Modern integrators can allocate peak heights in addition to peak areas, a measurement which previously could only be made from the recorder trace.

An integrator or computer has three tasks, which are to:-

1.   Allocate and measure areas or heights repeatably and accurately.

2.   Convince the chromatographer that in doing so it is using criteria with which he or she agrees, and

3.   Calculate compositions from raw area data according to an appropriate method of calibration.

Modern integrators work with the benefit of hindsight, and there is little doubt that the area or height is accurately measured provided that the beginnings and ends of the peaks are properly identified.   This is the significance of point 2 above, and on instruments which incorporate a printer/plotter, marks and symbols on the chromatogram which identify where integration starts and stops are a useful adjunct to the normal peak codes in the print-out.   Peak deconvolution (Littlewood, Gibb and Anderson, 1969) is not practicable on laboratory-sized machines and overlapping peaks are separated by dropping perpendiculars at valley points or by tangent skimming.   Modern instruments allow the areas to be reallocated without the chromatogram having to be rerun and the final results to be recalculated by a different method (see Chapter 8).   If a group of peaks are severely overlapped, so that neither dropping verticals at valleys nor tangent skimming is appropriate, then peak height measurements may still be reasonably accurate.

The recorder or plotter should create a faithful record of the chromatogram. This requires a fast response (less then a second for full-scale deflec-

tion), a high input impedance (greater than 10^5 ohms) and a small dead
band (less than 0.2% of scale). Most laboratory recorders are satisfactory
in this respect, and it is only for high-speed analysis, with peak widths of
less than one second, that other forms of recording the chromatogram may be
necessary.   On some integrators which incorporate plotters, the chromatogram is reconstructed from the digitised data, and the appearance of the
chromatogram can be affected by the integration parameters.   Thus if the
peak width value set into the integrator is too high, insufficient data
points will be accummulated for accurate reconstruction of the chromatogram.

Many newer chromatographs and integrators have auto zeroing facilities whose
operation should be treated with care since they are a potential source of
error.   Integrators will accept large positive signals (up to 1 volt), but
much smaller negative signals, designed to accommodate likely drift.   If
negative drift continues, and the plotter is automatically rezeroed at the
start of each chromatogram, it is easy to be unaware that the output signal
from the chromatograph is too low for the integrator to accept, and that the
true chromatographic baseline is lower than that presented on the plotter.
Fig 2.25 shows the effect.   Chromatogram A is recorded with the chromatograph output properly adjusted, whereas B, which may impress those with a
fondness for geometry, has the output offset beyond the range of the
integrator.   The solution is to suppress the autozero facility from time to
time and to check and where necessary adjust the position of the true
chromatographic baseline.

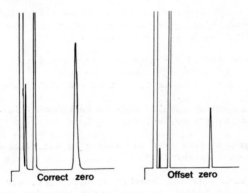

Fig. 2.25   Effect of incorrect zero on chromatogram.

CHAPTER 3

Separation in the Column

It has been pointed out (Chapter 1) that gas mixtures are very commonly separated by adsorption (gas-solid chromatography), so that operating temperatures considerably higher than the boiling points of the gases may be used. If heavier components in a mixture are also separated on an adsorption column, the operating temperature may need to be very high. All materials in the sample path, whether active solid, inert support, liquid phase, column and connecting tubing, valves etc. should ideally be compatible with sample components at the highest temperature at which they may be in contact with each other.

3.1   TUBE MATERIALS AND PHYSICAL FORM

The majority of gas analyses are performed on packed columns. Capillary columns, either coated or packed, have been used for particular problems such as separation of butane and butene isomers, but in general their high efficiency is not as necessary as in the petroleum or biological field. Since a variety of adsorbents is available, the more usual approach to a difficult problem has been to combine different adsorbents with one another, or with partition columns, using multi-column systems (see Chapter 5).

Stainless steel, aluminium and glass are the tubing materials most commonly used. Copper is compatible in many cases, but is not recommended where acetylene or sulphur compounds may be present. Aluminium or glass are preferred over stainless steel for low levels of sulphur compounds. P.T.F.E. has been used for trace levels of sulphur compounds and P.T.F.E. or nickel for corrosive gases. Where possible, plastics should be avoided, as air can diffuse through the tubing and, though undetectable, can degrade the packing material.

The shape of columns for gas analysis is subject to the same considerations which obtain generally in gas chromatography. Sharp bends should be avoided and so should any aspect of the shape which could cause voids to be left in the column packing, or fines to be created.

The diameter of columns is not usually critical in terms of the separation. The most common sizes are 2 mm i.d. (1/8" tubing) and 5 mm i.d. (1/4" tubing).   When using a concentration sensitive detector, such as thermal conductivity, the relationship between sample size and column diameter must be remembered.   For the same linear carrier gas velocity the volumetric flow rate through a 1/4" column will typically be 4-6 times greater than that through a 1/8" column.   To achieve a similar response, the sample size must be increased by this amount.   Conversely, with smaller diameter columns it may be difficult, when using a conventional gas sampling valve, to select a sample size which is small enough to prevent column or detector overload by a major component.

3.2   ADSORBENTS

3.2.1   Molecular Sieve

3.2.1.1   Zeolites   The name "molecular sieve" is rather general, but unless otherwise qualified is taken to refer to artificially prepared zeolites, which are the aluminosilicates of sodium, potassium, or calcium. Those most commonly used in gas chromatography are types 5A, calcium alumino-silicate, with an effective pore diameter of 5 angstrom units, and 13X, sodium alumino-silicate, with an effective pore diameter of 10 angstrom units.

The popularity of molecular sieves results from their unique ability to separate oxygen from nitrogen, using columns of normal length (1-2 m) at commonly used operating temperatures (ambient - 100°C).   The materials are also useful for analysis of the light gases H_2, CH_4, CO, NO and the inert gases He, Ne, Ar, Kr and Xe.

Molecular sieves must be activated by heating to drive off the water which otherwise occupies the pores, and consequently they can slowly lose their separating power because of adsorption of water either from samples, from carrier gas, or from the atmosphere.   Carbon dioxide is also strongly adsorbed at normal operating temperatures, and contributes to temporary deactivation of the sieves.   Activation can be carried out by heating the sieve under vacuum before packing the column, or by heating the packed column in the oven of the chromatograph with dry carrier gas flowing through it.

Janak, Krejci and Dubsky (1958), studying 5A molecular sieve, noted that the relative retentions of light gases, and particularly the order of elution of carbon monoxide and methane, depend strongly on the degree of activation. Aubeau, LeRoy and Champeix (1965), studying the same phenomenom, showed that with 9.4% residual water the order of elution of a series of gases was hydrogen, oxygen + nitrogen, carbon monoxide, methane + krypton and xenon. With 1.5% residual water, the order changed to hydrogen, oxygen, nitrogen + krypton, methane, carbon monoxide and xenon.   In between these, a water content could be found which allowed separation of all components.

Argon and oxygen are not separated under normal conditions.   Karlsson (1966) showed that when the column is activated in situ, the carrier gas used affects the subsequent separation.   2 metre columns of both 5A and 13X were activated with argon carrier at 450-500°C for 30 hours.   This gave good argon/oxygen separation in both cases, even though the argon carrier had contained 100 ppm water.   If helium was used during the activation, the water content had to be reduced to 10 ppm to achieve the same result.

Furthermore, the separation on sieves activated in helium deteriorated with use, whereas those activated in argon maintained their performance over long periods.

Pretreatment of molecular sieve is necessary for successful analysis of nitric oxide. Dietz (1968) and Clay and Lynn (1975) developed a procedure wherein the column was saturated with nitric oxide, which was then oxidised to nitrogen dioxide. This effectively neutralised the active sites on the sieve, allowing symmetrical peaks for nitric oxide.

Since both types of sieve are used for the same range of gases, the reason for choosing one or the other often depends upon the tactics of separation. Some of the features which might influence the decision are listed below.

5A   When fully activated, this provides the best separations, and would be preferred for difficult mixtures such as helium, neon and hydrogen. The long gap between methane and carbon monoxide, however, may be a disadvantage. Some large molecules are subject to exclusion phenomena on this material: sulphur hexafluoride passes through apparently unadsorbed, more importantly before oxygen (Simmons and co-workers, 1972), and iso-butane (2-methyl propane) elutes rapidly, in positions which vary with temperature and activity.

13X   Hydrogen, oxygen, nitrogen, methane and carbon monoxide are separated at fairly regular intervals. If it is desired to have carbon monoxide eluting before methane, this sieve would be the better choice, as at this degree of activation its long-term stability is excellent (Deans, Huckle and Peterson, 1971).

Figures 3.1 to 3.6 show the effect of increased amounts of activation on a 2 metre column packed with 5A molecular sieve. No separation is visible when the column is first packed with sieve which has been held in stock for some while, and the improvement with successively more intense activations is obvious. Because of the long retention time of carbon monoxide, Fig. 3.6 is recorded at half the chart speed of its predecessors.

The unusual behaviour of isobutane is illustrated by Figs. 3.7 and 3.8, where the column has been activated for 18 hours at 150°C and 250°C respectively. All chromatograms were recorded at 50°C.

Figures 3.9 to 3.13 show the effect of activation on a 2 metre column of 13X sieve. The gradual shift of carbon monoxide relative to methane is clear, and also the more regular spacing of components.

Batches of molecular sieve vary significantly in their performance, probably due to contamination of either type with the other. The chromatograms shown here illustrate the differences between the types, but should not be taken as typical. Both the maximum efficiency and the order of elution after a particular activation treatment may vary significantly from those shown.

A further unique property of 13X molecular sieve has been described by Brunnock and Luke (1968). Sieve treated with sodium hydroxide is packed into a 1 metre column and operated between 180 and 425°C. This gives separation of C_5-C_{10} hydrocarbons into naphthenes (cycloparaffins) and paraffins by carbon number. All the isomers of each compound type at each carbon number are contained in either a single peak (at lower carbon numbers) or a simple group of peaks (at higher carbon numbers). Although intended for naphthas and similar fractions, the method can be applied to gases, such as natural gas, by lowering the starting temperature. Separation of C_1-C_4 hydrocarbons presents no problem.

Fig. 3.1.   Chromatogram on unactivated molecular sieve, type 5A.

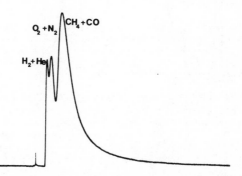

Fig. 3.2.   Chromatogram on molecular sieve, type 5A, activated for 3h. at 150°C.

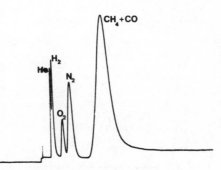

Fig. 3.3.   Chromatogram on molecular sieve, type 5A, actived for 18h. at 150°C.

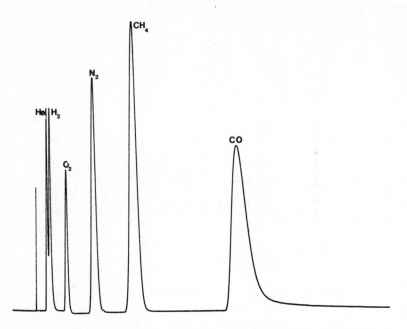

Fig. 3.4.   Chromatogram on molecular sieve, type 5A, activated for
3h. at 250°C.

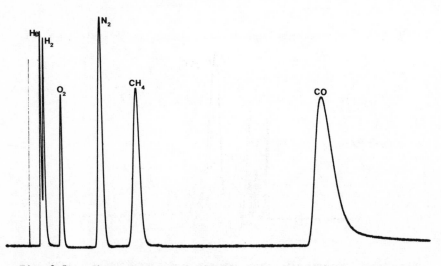

Fig. 3.5.   Chromatogram on molecular sieve, type 5A, activated for
18h. at 250°C.

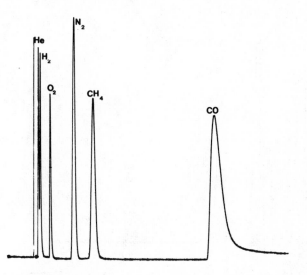

Fig. 3.6.   Chromatogram on molecular sieve, type 5A, activated for
18h. at 350°C.

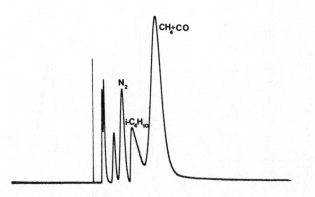

Fig. 3.7.   Chromatogram on molecular sieve, type 5A, activated for
18h. at 150°C, showing isobutane

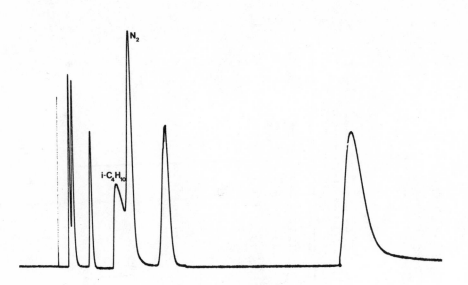

Fig. 3.8.   Chromatogram on molecular sieve, type 5A, activated for
18h. at 250°C, showing isobutane

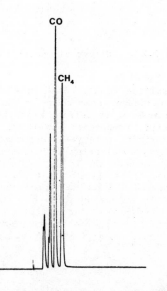

Fig. 3.9.   Chromatogram on molecular sieve, type 13X, activated for
3h. at 150°C.

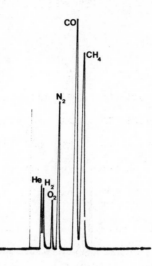

Fig. 3.10.   Chromatogram on molecular sieve, type 13X, activated for
18h. at 150°C.

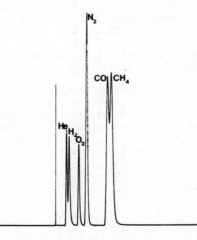

Fig. 3.11.   Chromatogram on molecular sieve, type 13X, activated for
3h. at 250°C.

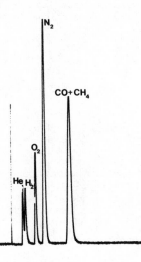

Fig. 3.12.   Chromatogram on molecular sieve, type 13X, activated for
18h. at 250°C.

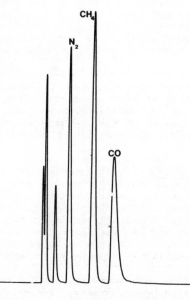

Fig. 3.13.   Chromatogram on molecular sieve, type 13X, activated for
18h. at 350°C.

3.2.1.2   Organic molecular sieve   Another types of molecular sieve has
been described by Riederer and Sawodny (1979). This is a polymeric Shiff-
base complex incorporating chromium, and has a pore diameter of 7 angstrom
units.   A 2 metre column packed with this material, and activated at
250°C, completely resolves the spin isomers of hydrogen when operated at
23°C.   All other reported separations have required temperatures substantially below ambient.

3.2.2   Silica

3.2.2.1   Silica gel   This material has been used since the beginning of
gas chromatography.   Janak (1953), describing his azotometer chromatograph, mentions columns of silica gel and active carbon.   When used with
temperature programming (Greene and Pust, 1957), it has separated hydrogen,
air, carbon monoxide, carbon dioxide and C_1-C_4 alkanes and alkenes.

The most common use in gas analysis has been as a complementary column to
molecular sieve for lighter components.   When operated at the same isothermal temperature, they allow separation of hydrogen, oxygen, nitrogen,
methane, carbon monoxide (on molecular sieve) and ethane, carbon dioxide and
ethene (on silica gel).   It has been largely superceded in this role by
porous polymer beads, but there may still be instances where its particular
characteristics may be useful, particularly in multi-column systems.   The
first group of components elutes rapidly in the order hydrogen, air, carbon
monoxide and methane.   There is then quite a long delay before ethane,
carbon dioxide and ethene emerge in that order.   This allows time, for
example, to switch the first group onto a molecular sieve column, on which
they can be completely separated before ethane elutes from the silica gel.

Fish, Franklin and Pollard (1963) used silica gel for combustion products,
which contained the gases described above and phosgene, hydrogen chloride
and chlorine.   Thornsbury (1971) described its use for Claus plant gases.
Because of variations between batches, a carefully prepared type (Deactigel)
was used.   After acid washing a 2 foot column at 120°C separates air,
carbon dioxide, carbonyl sulphide, hydrogen sulphide, carbon disulphide and
sulphur dioxide.

Silica gel adsorbs water strongly, and its chromatographic performance is
influenced by the amount of water adsorbed.   Like molecular sieve, it must
be activated before use by heating, although temperatures in the region of
150°C are usually adequate.   For the same reason, its performance gradually alters with use due to adsorption of water from various sources.

3.2.2.2   Porous silica   These materials are available as Spherosil or
Porasil, in a variety of surface areas and pore diameters.   The larger
surface areas and smaller pore diameters are used for gas analysis.
Guillemin and co-workers (1971a, 1971b), and Cirendini and co-workers (1973)
have studied the use of coated and uncoated porous silica.   Separation of
butane/butene mixtures on uncoated beads gives unsymmetrical peaks for the
unsaturated hydrocarbons.   Addition of water to the carrier gas modifies
the performance, giving faster analysis with symmetrical peaks.

Use of humidified carrier gas works because the water is adsorbed onto the
active sites which cause peak asymmetry, and maintained at an equilibirium
level.   An alternative approach is to modify the surface by coating a
conventional liquid phase onto the beads before packing the column.   As the

film thickness of the modifying liquid is increased, retention times reach a minimum and then slowly increase.   This reflects the fact that gas/liquid partition is taking over from gas/solid adsorption.   Peaks become almost completely symmetrical when their retention time is at this minimum and maintain their symmetry with increasing film thickness.

3.2.2.3   Durapak   An alternative approach to modification of the surface of porous silica   is chemical bonding of organic molecules to the surface (Little and co-workers, 1970).   Silica of intermediate surface area is used, and n-octane, oxypropionitrile, carbowax and phenyl isocyanate have been bonded to it.   The principal effect is still that of the silica, as is evident from the fact that n-octane/Durapak is more polar than carbowax/Durapak.   Presumably the bonded carbowax more effectively occupies the silanol -OH groups on the silica surface.

3.2.3   Alumina

Scott (1959) studied alumina deactivated with water for analysis of C_1-C_5 hydrocarbons.   The polarity, as measured by the retention of ethene relative to ethane and propane, reaches a minimum when the water content of the alumina corresponds to a monolayer coverage (about 2% w/w).   The level of activity is maintained by adding water to the carrier gas.   Deactivation with 2% silicone oil and 2% water allows separation of C_1-C_4 alkanes and alkenes, acetylene and the pentanes, except that iso-butene and trans-butene-2 coelute.

In one of the few instances of the use of a capillary column for gas analysis, Halasz and Heine (1962) described a packed   glass capillary containing alumina.   The degree of activity was adjusted by passing the carrier gas over sodium sulphate decahydrate.   McTaggart, Miller and Pearce (1968) evaluated a similar system for quantitative accuracy.   The 40 foot packed capillary gives very efficient separations, including all the butenes, in less than 12 minutes.

Al-Thamir, Laub and Purnell (1977) used alumina to attempt complete but slow separation of all C_1-C_4 hydrocarbons.   They modified the surface by coating with conventional liquid phases, and found that the separation varied not only with the amount and nature of the liquid phase, but also with the nature of the solvent used to disperse the phase.   The latter variation occurred even if the alumina was washed with the solvent alone and then dried before being packed.

More generally, however, for hydrocarbon analysis alumina has tended to be superceded by Durapak or graphitised carbon.   It is probably only the problem of maintaining constant activity that has caused its loss of popularity, and it should always be considered as a candidate for C_1-C_5 saturated and unsaturated hydrocarbons.

Alumina has been used for separation of spin isomers of hydrogen (Moore and Ward, 1968), and a column modified with $Fe(OH)_3$ has been used for hydrogen isotopes (Genty and Schott, 1970).   Both these separations were performed at $-196^{\circ}C$.

3.2.4   Carbon

3.2.4.1   Active carbon   This material has, like silica gel, been used
since the beginning of chromatographic gas analysis (Janak, 1953).   Ray
(1954) showed the separation of hydrogen, carbon monoxide and methane at
$20^{\circ}C$, and of hydrogen, methane, carbon dioxide, ethene, ethane and acetylene at $40^{\circ}C$.   Madison (1958) used a 25 foot column at $20^{\circ}C$ and separated hydrogen, oxygen, nitrogen, carbon monoxide and methane.   Since then
there has been little published on active carbon and its areas of application have been taken over by porous polymers.

3.2.4.2   Graphitised carbon   Graphitised carbon retains sample molecules
by geometrical structure and polarisability.   Modification of the surface
with relatively small amounts of liquids or solids produces significant
changes in performance.   DiCorcia and Samperi (1975) used graphitised
carbon for hydrocarbon gas analysis.   Modified with 1.5% carbowax, the C_2
hydrocarbons elute in the order acetylene, ethene, ethane.   Propene and
propane are resolved, and all the butanes, butenes and 1, 3-butadiene.   If
picric acid is used to modify the surface, unsaturates elute after saturated
hydrocarbons.

Bruner, Ciccioli and DiNardo (1975) used graphitised carbon + 0.7% phos-
phoric acid + 0.7% XE-60 for separation of hydrogen sulphide, sulphur
dioxide and methanethiol at parts per billion (1 in 10^9) concentrations.

3.2.4.3   Carbon molecular sieve   Pyrolysis of poly(vinylidene chloride)
produces a carbon with a large surface area which is inert and very non-
polar.   Kaiser (1970) has described its use for light gas analysis.   The
order of elution of hydrogen, oxygen + nitrogen, carbon monoxide, methane,
carbon dioxide followed by (in a temperature programmed run) acetylene,
ethene and ethane, is not very different from other forms of carbon.   Its
distinguishing features include high affinity for hydrocarbons.   This means
that water elutes before methane, and that rapid analysis of pentane requires temperatures which may pyrolyse the sample.

3.2.5   "Dusted" Columns

Column efficiencies in gas-solid chromatography are usually low, by comparison with those in gas-liquid chromatography.   Often, this represents no
problem as relative retentions of components are high, and they are easily
separated without high pressure drops and long analysis times.

In a gas-liquid column, the solid support particles are mainly responsible
for the gas phase diffusion term, and the liquid phase for the mass transfer
term (B and C respectively in the van Deemter equation).   In a gas-solid
column the adsorbent particles are responsible for both terms.   If the
adsorbent could be coated onto a conventional solid support the gas phase
diffusion term can be minimised as in gas liquid chromatography.   Bombaugh
(1963) demonstrated this effect by coating finely-divided molecular sieve
onto a diatomaceous earth support.   The packing contained 18% of the
adsorbent and had much greater efficiency than a conventional molecular
sieve column.

As an alternative to coating finely-divided adsorbent, Al-Thamir, Purnell
and Laub (1979, 1980) mixed normally-sized alumina with diatomaceous earth
support or with glass beads.   The columns generally contained 5-15%
alumina, and so the gas phase diffusion term was principally determined by

the inert support.   This allowed higher efficiency, and the relatively low content of adsorbent gave much reduced analysis time.

3.3   POROUS POLYMERS

The use of porous polymers as chromatographic separating media was introduced by Hollis (1966).   These materials are usually copolymers of divinylbenzene with another aromatic olefin and are used in the form of beads which are highly porous yet have a rigid structure.

They are used in many applications in preference to adsorbents and gas-liquid columns not only because of their special capabilities but because of their ease of use.   Columns can be prepared by simply pouring the material into the appropriate tube and "conditioning" (see below) according to the maker's instructions.   The resulting columns have a uniformity and certainty of performance superior to gas-liquid columns and considerably superior to that of adsorbents.

When used at temperature around 50°C, polymer bead columns will separate H_2, (O_2 + N_2 + Ar + NO + CO), CH_4, CO_2, N_2O and C_2H_6.   Subambient temperature allows separation of the O_2, N_2, Ar, NO and CO group.   At 100-150°C, C_3, C_4 and C_5 hydrocarbons can be analysed, and also H_2S, COS and SO_2.   Temperatures near the upper limit (c.240°C) are used for C_6, C_7 and C_8 hydrocarbons.

The exact mechanism of separation by polymer beads is not clear.

3.3.1   Commercial Polymer Beads

A number of commercial polymer beads are available and two groups widely used in gas anlyses are the Chromosorb "Century" Series (Johns-Manville Inc.) and the "Porapak" series (Waters Associates Ltd.).   The commercial beads contain some monomer and before use it is necessary to condition them by heating the column overnight at 250°C (190°C for Porapak T), with the carrier gas flowing and the column disconnected from the detector.

3.3.2   The "Century" Series

The types in this series are numbered 101 through to 108.

Chromosorb 102 is the one most used in gas analysis and is very similar to Porapak Q (see below).   Chromosorb 101 has similar separation character-istics to Chromosorb 102 but with reduced retention times;   its use therefore is more appropriate for high molecular weight materials rather than gases.   Chromosorb 103 is a polystyrene and has a special affinity for basic compounds; it therefore has a role in the analysis of gases containing ammonia and amines.   Chromosorb 104, 105, 106, 107 and 108 all differ in polarity from Chromosorb 102 and find use in exploiting certain separation and elution orders such as the examples described below.

3.3.3   The "Porapak" Series

Porapak Q is the one most widely used in gas analysis and is virtually interchangeable with Chromosorb 102.   Porapak P is similar in nature to Porapak Q but has a larger average pore size and hence lower retention

times, which makes it more appropriate to the analysis of high molecular
weight compounds.   Porapaks R, S, N and T are more polar than Porapak Q and
the resulting different separation characteristics are sometimes exploited
in gas analysis.   Types Q-S and P-S are silanised forms of types Q and P
respectively.

3.3.4   Separations by the Different Members of the Two Series

The small differences between the members of the series cause differences in
elution order of components of similar retention time and this allows
nuances of separation to be achieved by selecting appropriate members.
This can be particularly important in trace analysis since measuring a minor
component on the tail of a major component is quite difficult.   One example
of the differences between the behaviour of the different members is the
separation of propane, propene and water.   The following table shows the
order of elution at about ambient temperature.

TABLE 3.1   Elution Sequences on Different Bead Types

Bead Type	Propane	Propene	Water
Porapak P and Q Chromosorb 102 and 105	3	2	1
Chromosorb 101	3	1	2
Porapak R and S Chromosorb 103, 106, 107 and 108	2	1	3
Porapak T Chromosorb 104	1	2	3

These variations are such that an appropriate column can be chosen to elute
one of the components present as a trace amount before eluting the others as
major components.

A similar choice of order of elution is available for the commonly occuring
groups of components: ethane - ethene - acetylene and hydrogen sulphide -
sulphur dioxide - carbonyl sulphide.

3.3.5   Porapak Q and Chromosorb 102

These are probably the most widely used stationary phases in gas analysis.
This is because when first introduced they proved a simple replacement for
the relatively unpredictable silica gel for separating ethane, ethene and
carbon dioxide and subsequently because of their wide scope of application.

Their popularity is also enhanced by the almost exact coincidence between
the likely components of gas samples from whatever source, and the ability
of these materials to handle such samples within their operating temperature
limits.   At subambient temperature permanent gases can be separated, and
hydrocarbons up to C_8 have reasonably fast elution times at the maximum
operating temperature.   Heavier materials than C_8 can only be trace

components of gas mixtures and their measurement is a very specialised application. Furthermore, at these very low concentrations, their presence in a gas sample injected onto a porous polymer column is very unlikely to interfere with subsequent analyses.

Descriptions of applications are too numerous in the literature to cite here, though some specific ones are given in Chapter 6. For an overall view of their uses the continuously expanding applications notes available from the suppliers are recommended.

3.3.6   Others in the Two"Series"

Apart from Porapak Q and Chromosorb 102, others in the series do find occasional application to gas analysis for special separations. Some of these will be described in Chapter 6.

3.4   GAS LIQUID COLUMNS

As more types of adsorbent have become available, the use of liquid phases has declined. Separations which required long columns with high loadings of liquid phase can now be achieved by short columns of, for example, porous polymers. Furthermore, although Rein, Miville and Fainberg (1963) reported partial separation of oxygen and nitrogen on a fifty foot silicone oil column, gas liquid columns have not been used for permanent gases.

When light gases are present in a mixture, it is either accepted that they must be passed to an adsorption column (Isbell, 1963), or the absence of potentially interfering components must be ensured. Thus Beuerman and Meloan (1962) used dinonyl phthalate to separate mixtures which only contained oxygen, carbon dioxide and sulphur dioxide. In the same way, ASTM D 1945-64 recommends, among other choices, a 30 foot column packed with 27% silicone oil on Chromosorb P for analysis of natural gas. The success of this column in separating nitrogen, carbon dioxide and C_1-C_5 saturated hydrocarbons relies upon the absence of hydrogen, oxygen, argon, carbon monoxide and unsaturated hydrocarbons from any sample.

Propane/propene and butane/butene mixtures have been separated on liquid phases. I.P. 264/72 offers a large choice of columns for such uses, including a 30 foot column packed with Chromosorb P coated with benzyl cyanide and silver nitrate, as described in ASTM D 1717-65. This column relies on loose complex formation between unsaturated hydrocarbons and silver nitrate: ethene elutes after n-butane and propene after n-pentane. Packed columns coated with either di-n-butyl maleate or 1,8 - dicyano-octane on diatomaceous support can be used for the separation of C_3 and C_4 hydrocarbons.

A mixed bis-lactam phase (Ravey, 1978) has been used for C_3-C_5 saturated and unsaturated hydrocarbons. The phase was made by condensation of pyrrolidone and caprolactam with formaldehyde, and packed, as 10% on Chromosorb P, into a 10 metre column.

CHAPTER 4

Choice of Carrier Gas

4.1 INTRODUCTION

The influence of different carrier gases used for chromatographic gas analysis is subject to the same rules which apply generally in gas chromatography. Theory in the form of the van Deemter equation (van Deemter, Zuiderweg, and Klinkenberg, 1956), and practice both show that:-

(a)   a carrier gas of higher molecular weight can, at the minimum of the H/u curve give a more efficient separation than one of lower molecular weight and

(b)   a carrier gas of lower molecular weight displays the minimum of the H/u curve at a higher carrier gas velocity than one of higher molecular weight.

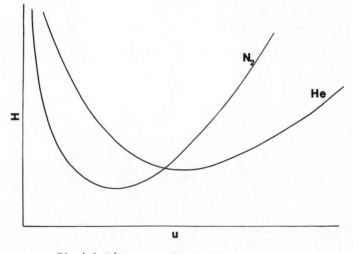

Fig 4.1 H/u curves for different carrier gases

This is illustrated schematically in Fig. 4.1.   Carrier gas velocity is a more easily altered parameter than column length, and consequently workers wishing to speed up an analysis will often use a column which is longer than strictly necessary at a flow rate higher than optimum.   It is evident from Fig 4.1 that low molecular weight gases such as helium have an advantage in this respect.

Over and above these general considerations the use of detectors which measure the difference in properties between carrier gas and eluted component, such as the thermal conductivity detector, influence the choice of carrier gas considerably.   When analysing gases, as opposed to liquids or solids, there is the possibility that the carrier gas is a component of the sample or that it has properties which are very similar to those of sample components.

4.2   THERMAL CONDUCTIVITY DETECTOR

The difference in thermal conductivity between carrier gas and component is not the only factor which determines detector response but it is usually the most significant.   The table of thermal conductivities (Table 4.1) indicates the approximate sensitivites to be expected when different carrier gases are used. Hydrogen and helium give high sensitivity with this detector, and allow more rapid analysis.   Of the two, helium is much preferred on safety grounds and also because it is a less common component of gas mixtures. When using helium as carrier, anomalous results can be obtained if hydrogen is a sample component (see 4.7.2).

TABLE 4.1   Thermal Conductivities of Gases (Weast, 1977a)

Gas	Thermal Conductivity	$\delta TC(He)$	$\delta TC(N_2)$	$\delta TC(Ar)$
Hydrogen	471.1	95.0	405.4	425.6
Helium	376.1	–	310.4	330.6
Methane	89.3	−286.8	23.6	43.8
Oxygen	68.2	−307.9	2.5	22.7
Nitrogen	65.7	−310.4	–	20.2
Carbon monoxide	63.9	−312.2	−1.8	18.4
Ethane	58.3	−317.8	−7.4	12.8
Ethene	55.0	−321.1	−10.7	9.5
Propane	48.4	−327.7	−17.3	2.9
Argon	45.5	−330.6	−20.2	–
Carbon dioxide	43.8	−332.3	−21.9	−1.7
n−Butane	43.4	−332.7	−22.3	−2.1

The variations in response which are to be expected when using different carrier gases can be used in a number of ways, as the following examples show:-

(i)   The measurement of oxygen in flue gases.   Flue gas contains a reduced level of oxygen, but a similar level of argon to that in the combustion air (0.93%).   Argon and oxygen are difficult to separate unless subambient temperature is used.   Helium would give high sensitivity, but argon/oxygen separation would be necessary.   If, however, argon is used as carrier gas the separation is not necessary.   The choice would be to use argon as carrier and to compensate for the low sensitivity by sample size.

(ii)   The analysis of a mixture of butanes and butenes.   If argon is used as carrier gas, some components will give positive peaks, some will give negative peaks and some will give virtually no peaks at all; simplistically this can be attributed to the thermal conductivity of argon lying within the thermal conductivities of the butanes and butenes.

With helium as carrier gas, the butanes and butenes give all positive (or all negative) peaks and the relative sensitivity for each component is approximately the same.   For this application helium would be the choice.

(iii)   The measurement of a low level of carbon dioxide in nitrogen.   If helium is used as carrier gas, a high sensitivity to both components is obtained;   however, the nitrogen peaks may tend to be broad with large samples, overlap the carbon dioxide peak and hence limit the available sensitivity in measuring the carbon dioxide.   If nitrogen is used, the intrinsic sensitivity is low but as there is no need to separate nitrogen from carbon dioxide, a large sample and a short column should be effective. In this case either helium or nitrogen can be used, the choice depending on any other requirements from the analysis.

4.3   FLAME IONISATION DETECTOR

Since the FID does not in normal operation respond to permanent gases the choice of carrier gas is much less influenced by detector characteristics. A gas of lower thermal conductivity, such as nitrogen, allows maximum response from the detector, as a result of higher flame temperature. Unless the analysis requires that the detector be operated at maximum sensitivity, helium will probably be the best choice for the advantage it offers in analysis time.

4.4   ELECTRON CAPTURE DETECTOR

The choice of carrier does affect the correct operation of the ECD.   Most commercial detectors work satisfactorily on nitrogen which is much more convenient than the alternatives of 5% or 10% methane in argon.

4.5   OTHER DETECTORS

Flame photometric detector.   The choice is similar to that for the flame ionisation detector.   However, the oxygen level in the carrier should be checked as oxygen can adversely affect the measurement of oxidisable sulphur compounds such as mercaptans.

Ultrasonic detector.   The response is a function of the difference between the molecular weight of carrier gas and sample component.   The choice of carrier gas will be controlled by the fact.

Helium ionisation detector.   For correct operation, the carrier gas must be ultrapure helium.   The possibility of carrier gas contamination by leakage of air into the system or of back-diffusion of air into the detector must be eliminated.

4.6   CARRIER GAS PURITY

This section deals with undesirable impurities, rather than components deliberately added to the carrier which are considered in 4.7.

Most carrier gases are available in high-pressure cylinders at purities which are adequate for many purposes.   A convenient shorthand is used by compressed gas suppliers to describe purity.   A two-digit code defines the number of 9s in the percentage purity value, and the final decimal.   Thus 99.6% is 2.6, 99.99% is 4.0 and 99.9995% is 5.5.   The likely impurities are other permanent gases, water vapour and light hydrocarbons.

The significance of particular impurities varies with the application, depending upon the sample, the column and the detector.   Sample components may be prone to oxidation, which could cause them to be lost (e.g. silane) or converted (thiols → disulphides).   Oxygen or water vapour can degrade a column packing which is being used at or near its upper temperature limit – porous polymers may be subject to this in gas analysis.   The performance of the more sensitive and specific detectors can be ruined by impurities to which other types of detectors would be completely oblivious.   A trace of intensely electron-capturing materials, which may be undetectable by any other method, can prevent proper operation of an ECD.   The trace need not come from the cylinder – the authors have suffered the effect of having cleaned carrier gas lines with halogenated solvent.

Deoxygenating and drying traps in the carrier gas line allow longer column life and better detector performance.   Oxygen is removed by transition metal suboxides and water by molecular sieves, which will also adsorb any higher molecular weight impurities.   Both types of trap must be regenerated or replaced periodically.   Both oxygen and water vapour can diffuse through nylon tubing and so carrier gas lines should be plumbed in metal when traps are considered to be necessary.   Recently, a device for removing water vapour by selective permeation through a tube bundle has become available. It has the advantage of continuous performance, but the authors have no experience of how it compares with molecular sieve dryers.

The helium ionisation detector, with its extreme sensitivity to permanent gases, is very much affected by trace impurities.   To produce the ultrapure helium which is needed, heated hopcalite for oxidation of hydrocarbons and carbon monoxide has been used with molecular sieve traps at liquid nitrogen temperature.

4.7   MIXED CARRIER GASES

Components may be added to the carrier gas at trace levels to improve column or detector performance, or sample components may be added at percentage levels.   These represent different types of application.

4.7.1   Trace Additives

The presence of small amounts of other components in the carrier gas can
affect column performance or detector characteristics.   In the former area,
the only applications in gas analysis concern addition of water vapour so as
to maintain the degree of activity of alumina or porous silica columns for
light hydrocarbons.   McTaggart, Miller & Pearce (1968) used an alumina
capillary column, with carrier gas passed through a tube packed with a
hydrated metal salt, such as $CuSO_4$ $5H_2O$.   This was maintained at a
constant temperature and hence provided a constant water vapour pressure.

The ECD has been shown to be sensitised to sample components to which it
would not normally respond by the addition of traces of oxygen or nitrous
oxide to the carrier (see section 2.5.2).   In this case, since the added
component does not in any way affect the column performance, it could be
added between the column and the detector.   Indeed, if one wished to try the
effect of varying the level of additive, or to switch between levels, this
would be the easier arrangement.   Furthermore, if the carrier gas were to be
split between two detectors, the other of which may be affected by the
additive, its addition just before the detector would avoid the problem.
The helium ionisation detector has been shown to behave more consistently
with controlled levels of other permanent gases in the carrier (see section
2.5.7), and these can also be added immediately before the detector.

4.7.2   Addition of Sample Component

In the 1960s vacancy chromatography, when using a thermal conductivity
detector, was a popular idea, although its popularity was perhaps more in
concept than in practice.   If the carrier gas were to contain a sample
component, at the anticipated or desired concentration, then injection of
the sample should show, by the size and direction of the signal, the nature
of the variation from the intended level.   Annino, Franko and Keller (1971)
showed that the practice is more complicated than this simple outline, since
the size of the signal for the component depends not only on the difference
between its concentration in the sample and in the carrier gas, but also on
the identities and concentrations of all other components.

One application in which mixed carrier gas is practicable, however, concerns
analysis of hydrogen with helium carrier gas.   Hydrogen gives a relatively
small response in helium, as would be expected from Table 4.1.   What would
not be anticipated, however, is that at low concentrations the signal is in
the same direction as that for all other components, and that as the concentration increases, the peak inverts at its maximum, eventually becoming
strongly negative.

The reason for this behaviour is due to the shape of the thermal conductivity curve, which is shown in Fig. 4.2.   The minimum occurs at about 6% of
hydrogen and the mixture with the same thermal conductivity as helium
contains about 12%.   These are higher concentrations than are normally
found at the end of the column, but hydrogen is barely retained on most
columns and can reach these levels.   The effect on the peak shape as the
concentration at the peak maximum reaches and exceeds these levels can be
seen from Fig. 4.2.

If a mixture contains hydrogen as a major component, it can be measured by
using argon carrier, but with significantly reduced sensitivity to other
components, or by using helium carrier and limiting the sample size so as

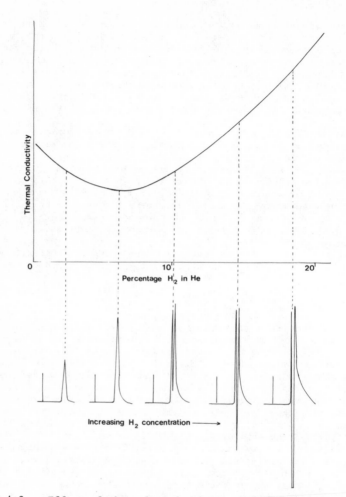

Fig. 4.2.   Effect of thermal conductivity of H_2-He mixtures on
            peak shape.

not to approach the point of peak inversion (Pauschmann, 1964), which again
reduces the net sensitivity to other components.   Alternatively, some
hydrogen can be mixed in with the carrier gas so that the concentration of
hydrogen in helium is always greater than that at the point of inversion.
This was proposed by Purcell and Ettre (1965), who used a commercially
available mixture of 8.5% H_2, 91.5% He.   This gives sensitivity for other
components similar to that found with pure helium and a peak of opposite
polarity for hydrogen with no inversion.

In the authors' experience, this mixture produces some non-linearity in the
calibration for hydrogen, since at 8.5% the thermal conductivity curve is
close to the minimum, and the thermal conductivity of the mixture is not yet
a linear function of the extra concentration of hydrogen.   If the hydrogen
content is increased to 20% a linear calibration plot is obtained.   The

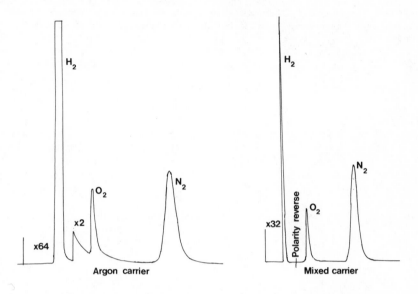

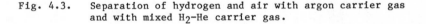

Fig. 4.3.   Separation of hydrogen and air with argon carrier gas
and with mixed H_2-He carrier gas.

advantage of using this mixture is evident from Fig. 4.3, which shows the
separation of a 90% hydrogen, 10% air mixture on molecular sieve. Chromatogram A is obtained with argon carrier and B with hydrogen/helium carrier.
The mixed carrier gas requires a polarity change, but allows more accurate
measurement of other components. An interesting side-effect concerns the
relative retentions of helium and hydrogen. Figure 4.4 chromatogram A shows
the separation of the two gases on a molecular sieve column using argon
carrier. When the same column is used with mixed carrier gas, chromatogram
B shows that helium and hydrogen have apparently identical retention times.
Presumably the sites which adsorb these gases are fully saturated from the
carrier gas - no adverse effect on separation of other components is
observed, however.

It has been suggested from time to time that this mixed carrier gas is only
usable if the concentration of hydrogen in the sample is greater than that
in the carrier, and that lower concentrations will give negative (vacancy)
peaks. This can occur only if the other component in the sample is helium,
or if the H_2:He ratio in the sample is less that that in the carrier. In
this case the negative hydrogen peak is in fact a positive helium peak. In
the absence of helium, the following argument shows that the response is
consistent.

As the hydrogen peak elutes from the column, the total number of moles in
the detector cell does not change. Since the cell now contains an amount of
hydrogen from the sample it must contain a reduced amount of carrier gas.
The only components present in the cell are hydrogen, from both sample and
carrier, and helium from carrier alone. As the amount of carrier gas is

reduced, the helium concentration is reduced, and so the hydrogen concentration is increased.   This is true irrespective of the amount of hydrogen
eluting from the column, and hence of the concentration of hydrogen in the
sample.

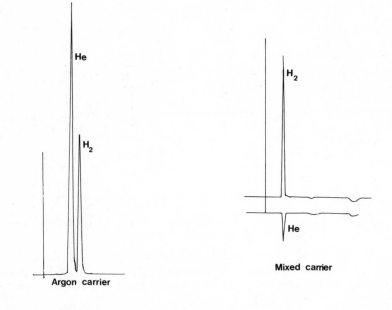

Fig. 4.4.   Separation of hydrogen and helium with mixed carrier
             gas.

CHAPTER 5

Tactics for the Analysis of Gas Mixtures

5.1 INTRODUCTION

The analyst is usually required to measure a number of components in a gas mixture, and a single column in a single isothermal chromatograph will rarely meet this need. A mixture often contains components whose separation dictates a particular column or set of conditions and also components whose measurement is not possible with this column or these conditions. The most frequent example concerns mixtures containing air and carbon dioxide (and possibly water vapour). Molecular sieve is needed for oxygen/nitrogen separation at normally used temperatures (ambient to $100^{\circ}C$); carbon dioxide has an extremely long retention time and water vapour is adsorbed onto the sieve under these conditions. Conversely, columns suitable for carbon dioxide or water measurement, such as porous polymer beads, will not under the same conditions separate air components.

The use of a number of chromatographs can offer a solution but may be uneconomic. Fortunately the analyst can use a range of tactics, singly or in combination, to extend the scope of a single chromatograph. A number of these tactics are described below and examples of their application will be found in the next chapter.

5.2 ISOTHERMAL ANALYSIS

Temperature has a dramatic effect on the retention and separation of components of a mixture on a column and it is therefore considered as the first tactical variable. If a complex mixture is separated on a well chosen column at a temperature of, say, $70^{\circ}C$, one finds that there is a group of components which are conveniently separated for quantitative estimation. This is preceded by a group of components which are inadequately resolved and followed by another group for which the peaks are low, broad and of excessive retention time, as in Fig. 5.1.

Lowering the column temperature to, say, $40^{\circ}C$ will improve the separation of the earlier components but will cause broadening of the previously estimable components. Raising the temperature to, say, $150^{\circ}C$ will reduce

61

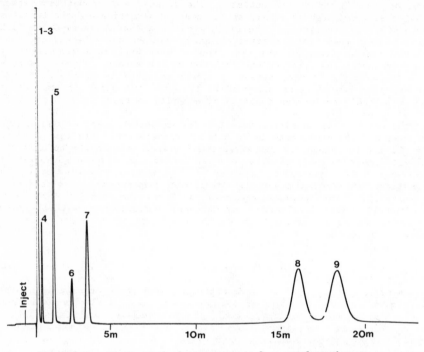

Fig. 5.1   Isothermal chromatogram of a complex mixture

the retention time of the latter components, sharpen their peaks and make them estimable.   The previously estimable components will not be resolved.

A single chromatograph used with one analytical column can, by changing its oven temperature, resolve and estimate a wide range of components.   A prerequisite of the successful performance of many of the other tactics described below is the selection of an oven temperature such that one or more columns will give suitable resolution of a selected group or groups of components.

5.3   THE USE OF TWO COLUMNS WITH TWO DETECTORS

5.3.1   Series Operation

The selectivity of different columns, and the fact that speed of analysis and separation both vary on the same column when its temperature is varied, led to the suggestion of using columns in series.   In the simplest form (Davis & Schrieber, 1957) two columns are used.   The first is designed for the heavier components of the mixture and hence rapidly elutes the lighter ones substantially unresolved on to the second column which is able to

separate these light components.   A detector is placed at the junction of
the two columns and at the outlet of the second.   A conventional thermal
conductivity detector is ideal, as it is non-destructive and the two sensors
can act as the detectors.   The first signal is from the unresolved lighter
group as they pass from the first column to the second.   Depending upon the
relative performances of the two columns, one will then see the heavier
components from the first column followed by the resolved lighter components
from the second, or vice versa.   If the mixture is complex, it may be
necessary to record peaks alternately from the two sides of the detector,
which is likely to involve a series of polarity changes.

The two columns can be fitted into the chromatograph oven, or one can be in
the oven and the other outside at ambient temperature.   In the latter case
the ability to change the temperature difference between the columns should
make it simple to avoid simultaneous elution of components.   Even with both
columns in the one oven, changes in temperature should still allow
adjustment of elution order, as different columns react to temperature
differently.   More complex applications (Terry and Futrell, 1965, Boreham
and Marhoff, 1960) involve multiple columns and detectors with the signals
from the detectors being switched sequentially to the recorder at the
appropriate times.

5.3.2   Parallel Operation

If the injected sample is fed to a tee-piece to which two columns (of
dissimilar characteristics) are connected, it will be split between them.
The outlets of the columns can then be reconnected before the detector, or
fed to the two sides of a thermal conductivity detector, or to two independent detectors.   A particular example involves the use of a column
consisting of two concentric tubes.   The inner tube contains one packing
and the annular space contains the other (Alltech Ass. 1978).

Elution of components is similar to that for series operation, except that
the lighter components have not travelled through both columns.   There is
no guarantee that a pair of columns which give good separation in series
will do so when connected in parallel, or vice versa : the difference in
behaviour is principally due to carrier gas compressibility which affects
the column performance differently in the different modes.

One would assume that the sample might split between the columns in the same
ratio as does the carrier gas.   This may not be true, especially if the
viscosities of carrier gas and sample are significantly different.   More
importantly, the viscosities of different samples may vary and so the split
ratio may not be consistent.

A disadvantage of either method of working is that the heavier components do
go onto the column intended for the lighter ones.   This means that the time
between analyses is extended, unless the potential interference is so small
that it can be ignored.

The method's advantage is that it requires very little modification to
commercial equipment.

5.4   SWITCHING VALVES

The use of low dead-volume valves in the carrier gas flow path allows the analyst to take wider advantage of the characteristics of different columns. The applications listed below are not exhaustive, but indicate the more common areas of use.   Where a particular valve is shown, the same application can be performed by a valve with more ports : it may be necessary to link unused ports, or to blank them off.

Valves can be used for:

> Column selection
>
> Backflushing
>
> Column isolation
>
> Multiple functions, combining the above
> with each other or with gas sampling

Four-port and six-port valves can perform most of the operations above. Eight-port and ten-port valves are used where more than one function is desired from a single valve.

5.4.1   Configuration of Valves

Schematic drawings of valves differ, since manufacturer's designs vary as regards the orientation of the ports in the valve, the positions of the slots in the rotor which connect pairs of ports, and the angle of rotation. For simplicity, the diagrams in this chapter follow the convention, where possible, that the ports are equally spaced at a constant radius from the axis of rotation of the valve.   Where alternative schematics are commonly used, the equivalence is explained by reference to the following figures.

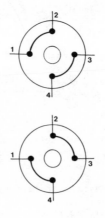

Fig. 5.2.   Four-port valve

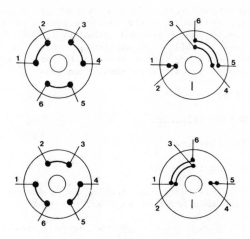

Fig. 5.3. Six-port valves, two formats

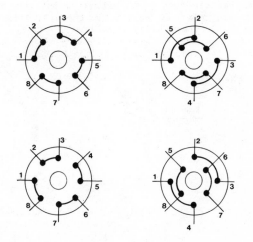

Fig. 5.4.   Eight-port valves, two formats

For many applications these eight-port valves are interchangeable, but there are particular uses which one format and not the other will satisfy.   The valve on the left turns through 45° and that on the right through 90°. They are distinguished by this in the subsequent text.

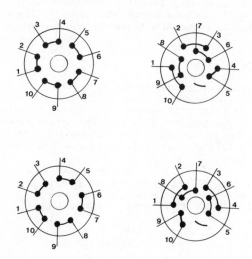

Fig. 5.5.   Ten-port valves, two formats

5.4.2   Column Selection

A four-port valve is used as shown in Fig. 5.6.   Operation of the valve
allows the injected sample to go onto either column.   The two detectors
could be the sensor and reference cells of a thermal conductivity detector.

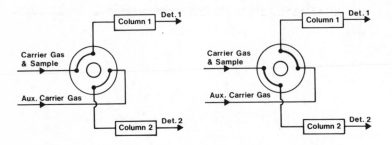

Fig. 5.6.   Column selection

When used in this way the system may not offer any advantage over the use of
another gas sampling valve.   With a gas sampling valve in each carrier gas
line, samples can be injected onto either column without the baseline
disturbances which may attend the use of a switching valve.

If the effluent from one of the columns is to be switched to different detectors during an analysis, then the configuration of Fig. 5.6 can be modified by fitting the columns before the valve. The effluents from the two columns can then be switched between the two detectors.

5.4.3   Backflushing

If the sample is complex, optimising the separation of a group of components may mean that heavier components have very long retention times. Conditions may not favour their measurement and they can cause an excessive delay between analyses. The gas flow in the column can be reversed after elution of the last component of interest, and the heavier components eluted backwards : this should take approximately as long as they have spent travelling forwards in the column.

The backflushed components can be vented, or fed to the detector so that a total figure for heavy components may be measured. While components are being backflushed, they are to some extent regrouped : those components which travel faster in forward flow also travel faster in backflush, and so catch up with the slower-moving ones. This is a somewhat simplified view and depends on the pressure drop and the ratio of retention times of the heavier components.

Peak broadening cannot be reversed, but the backflushed components often appear as a single, perhaps rather unsymmetrical peak. The limit of detection of this group is obviously much lower than that of the individual components which comprise it.

5.4.3.1   Backflush column to detector.   A four-port valve is used as shown in Fig. 5.7.

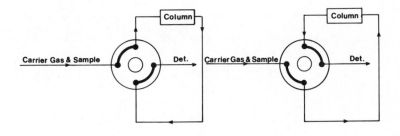

Fig. 5.7. Backflush column to detector

If the detector is an FID, the pressure surge when the flow is reversed may extinguish the flame: a restrictor should be fitted between the valve and the detector.

5.4.3.2   Backflush pre-column.   With the above configuration, the total time taken is approximately twice that of the forward flow. An alternative approach uses two columns, the first of which (pre-column) need only be long enough to separate the group of desired components from the group to be backflushed. The first group is passed to the second (analytical) column

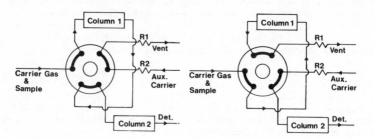

Fig. 5.8.   Backflush pre-column to vent

in which they are separated.   Simultaneously, the second group is back-flushed from the first column.   A six-port valve is used as shown in Fig. 5.8.

Restrictors R1 and R2 are set to have similar pneumatic resistances to the analytical column and the pre-column respectively.   This should ensure that the flow through the analytical column is undisturbed by the operation of the valve.

The backflushed group of components can be measured, if desired, by placing another detector on the vent line.   Alternatively the vent line can be teed in to the existing detector after the analytical column.   In this case the column lengths and switching time must be chosen so as to avoid interference between the backflushed group and the normally eluting components.

5.4.3.3   Backflush and sequence reverse.   A six-port valve is used with a pre-column and an analytical column.   A single carrier gas source is used, and the plumbing arranged so that switching both reverses the flow through the pre-column and positions it after the analytical column.   Figure 5.9 shows the configuration.

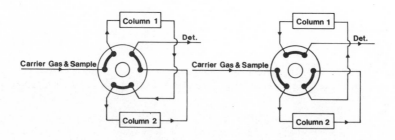

Fig. 5.9.   Backflush and sequence reverse

The pre-column would normally be chosen to be considerably shorter than the analytical column.   This means that it has little effect on the normally eluted components, even though they pass through it twice.   The backflushed group of components, however, having spent only a short time travelling in either direction will emerge as a sharp peak.   It can be arranged for the backflushed group to be the first to elute : this allows much lower detection limits than with a more conventional broad backflushed peak.

5.4.4   Column Isolation

Series operation of columns allows measurement of groups of components separated on columns whose characteristics are different.   The drawback of the technique is that all the components eluted from the first column pass to the second.   This can mean excessive retention times, which prolong the interval between analyses, or deactivation of the second column (e.g. H_2O, CO_2 on molecular sieve).

A switching valve can be used to isolate and by-pass the second column after the appropriate components have entered it.   Elution proceeds from the first column directly to the detector, then subsequently the second column is reconnected and the components which have been stored in it are passed to the detector.   Alternatively, it may be possible to continue elution through the columns in series until all components from the second column have been measured, then by-pass the second column before the heavier components have reached it.

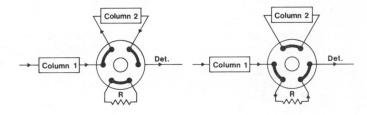

Fig. 5.10.   Column isolation.   Method 1

Figure 5.10 shows the use of a conventional six-port valve.   The restrictor is adjusted to have the same resistance as the column, so that the flow is independent of the valve position.   A disadvantage of the configuration shown in Fig. 5.10 arises from the fact that both ends of the column are connected via the valve when it is isolated.   During isolation, the pressure of the carrier gas in the column will adopt a value intermediate between the inlet and outlet pressures under flow conditions.   This is achieved two ways : by continuing forward flow through the column, and by reverse flow from the high-pressure to the low-pressure end of the column via the valve.   There is a danger that a component which has just entered column 2 before it is isolated may be transferred after isolation to the low pressure end of the column.   This is avoided by the configuration shown in Fig. 5.11.

Here the use of a rotor with only two slots means that equalisation of pressure in the isolated column is only achieved by forward flow of carrier gas.

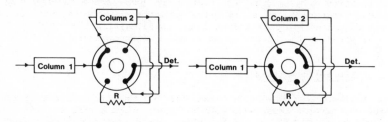

Fig. 5.11.   Column isolation.   Method 2

When components are isolated in a column in this way, the chromatographic process stops, but the component bands are still subject to diffusion in the mobile phase, which causes some broadening.   The degree of broadening depends upon the time and upon the partition coefficient of the component : significant band broadening only occurs in the mobile phase and hence affects lighter components to a greater extent.   In general, within the time-scale of most analyses, the degree of broadening will be perceptible but not troublesome.   Furthermore, this type of operation will almost invariably be controlled by an automatic timer, and so the peak shape and resolution will be consistent.

5.4.5   Foreflushing

Column isolation allows maximum flexibility when multiple columns are being used, as components can be stored until the most appropriate time in the analytical cycle.   There is the danger, however, that if the isolating valve is not completely leak-tight, stored components may leak out or small amounts of air may leak in.   Foreflushing has been described by Willis (1978) as a way of avoiding this.   Figure 5.12 shows the use of a six-port valve for this purpose.

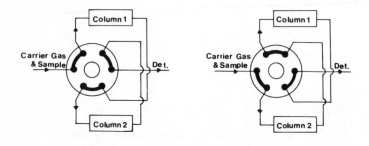

Fig. 5.12.   Foreflushing

After sample injection, and when the lighter components have been passed on to column 2, the valve is rotated, and the heavier components eluted directly from the column 1 to the detector.   The lighter components, having been separated on column 2, now pass through column 1 for a second time and are then detected.   Provided that they have been sufficiently well separated on column 2, the small extra amount of peak broadening caused by this second passage through column 1 is not significant.   Column lengths must be selected so that there is no overlap of components.

5.4.6   Multiple Functions

When an analysis requires more than one of the above operations, or a combination of any of these with sample injection, a series of valves can be used.   Alternatively where two functions can occur at the same time, they can be dealt with by a single multi-port valve.   Some typical examples follow.

5.4.6.1   Gas sampling and backflush to detector.   An eight-port valve with 45° rotation is used as shown in Fig. 5.13.   The left hand position shows the sample being loaded into the loop.   The right hand position shows the sample being injected and flowing through the column in what would be considered to be the forward direction.   Components separated in this way continue on to further columns or to the detector.   When the valve is returned to the left hand position the column is backflushed and the sample loop returned to the load position.

The timing of the backflush operation is dictated by the requirements of the analysis.   The time at which the sample loop is returned to the load position is not critical, and so the fact that these operations necessarily occur simultaneously is not a disadvantage.

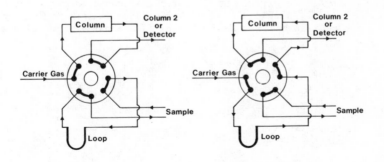

Fig. 5.13.   Gas sampling and backflush to detector

5.4.6.2   <u>Gas sampling and backflush to vent.</u> A ten-port valve is used with
an auxiliary supply of carrier gas as shown in Fig. 5.14.

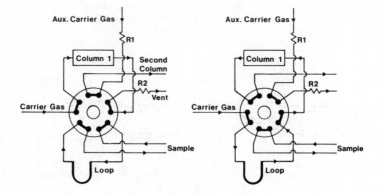

Fig. 5.14.   Gas sampling and backflush to vent

Restrictor 1 is adjusted to have the same resistance as Column 1, and
restrictor 2 to the same resistance as any downstream columns.

5.4.6.3   <u>Gas sampling and sequence reverse</u>   A ten-port valve combines
these functions as shown in Fig. 5.15

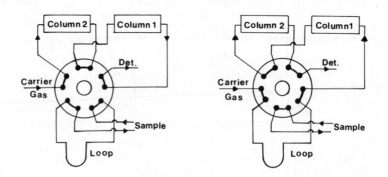

Fig. 5.15.   Gas sampling and sequence reverse

5.4.6.4   Column selection and backflush.   Either of two columns is
selected : the other is simultaneously backflushed.   An eight-port valve
with 90° rotation is used as shown in Fig. 5.16.

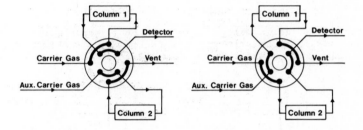

Fig 5.16.   Column selection and backflush

5.4.7   Advantages and Disadvantages of Valves

The above examples show some of the ways in which valves add considerable
extra flexibility to chromatography.   Valves are now designed with sufficiently small dead volumes so that they do not   perceptibly affect the
quality of separation when used with packed columns.   They have even been
shown (Miller, Stearns and Freeman, 1979) to be useable with capillary
columns.

Valves, particularly with automatic actuators, are fairly expensive, although their use should save time and money elsewhere.   Alterations must be
made to the chromatograph in terms of plumbing and mounting the valves.
Most modern chromatographs are designed with this in mind.

They are conventionally fitted inside the column oven or in an adjacent
auxiliary oven.   In the former case they limit the maximum operating
temperature of the oven to values which depend upon the rotor material.
When analysing gases free from large amounts of condensible material, it is
possible to use valves external to the oven.   If the original sample was a
stable gas at ambient temperature then the individual components when
further diluted with carrier gas should not be subject to condensation on
surfaces at ambient temperature.   Use of valves externally in this way no
longer limits the column oven temperature, but requires longer connecting
tubing, which may begin to affect the separation.

Corrosive or highly adsorbent sample components may not be compatible with
the materials in the valve.   Valve bodies are usually stainless steel, and
the rotor is filled P.T.F.E. or polyimide.   Valves with pure tantalum
bodies and connecting tubing are available, which can be connected to glass
columns by gold-plated fittings and polyimide or graphite ferrules.

5.5   PRESSURE BALANCING

Carrier gas flows through a column at a rate and in a direction controlled
by the applied pressure.   Switching valves organise the flow routes so that
the applied pressure works in the desired way.   If, on the other hand,
carrier gas pressure can be externally controlled and applied to the point
in the chromatographic system where a change in flow is required, then
switching valves in the flow path are not necessary.

This principle of using externally controlled pressures was devised by
Deans, and has been described for backflushing, heart-cutting, column
selection, peak storage and adjustment of the separation characteristics of
columns used in series.

5.5.1   Backflush to Vent (Deans, 1965)

As in 5.4.3.2 two columns are used in series.   The first, or pre-column,
need only separate the components of interest as a group from the heavier
components.   Separation of the first group continues on the second, or
analytical, column while the heavier group are backflushed from the pre-
column.   Figure 5.17 shows the layout.

Modifications required inside the column oven are the addition of two
T-pieces : the one fitted between the columns should be of low dead-volume
relative to the flowrate of carrier gas used.   Modifications external to
the oven involve the addition of a second pressure regulator, PR2, and
gauge, PG2, two on-off taps (commonly solenoid valves) A and B, whose
operation is simultaneous but opposite, a fixed restrictor R1 and an
adjustable restrictor R2.

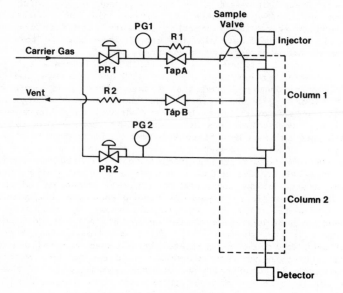

Fig. 5.17.   Backflush to vent

To set up forward flow Tap A is open and Tap B closed.   Pressure regulator
PR1 is adjusted to give the required flow through both columns.   Pressure
gauge PG2 shows the natural junction pressure under these conditions.   PR2
is now adjusted so that PG2 reads a slightly higher pressure.   This means
that most of the flow through column 2 is supplied via column 1, but a small
additional make-up flow enters at the T-piece via PR2.

To backflush, Tap A is closed and Tap B opened.   PR2 now supplies all the
flow to column 2, at the same rate as before, since the pressure at the
junction has not changed, and also supplies carrier gas in the reverse
direction through column 1, Tap B and R2 to vent.   R2 is adjusted so that
the backflush flow is similar to the forward flow : the setting is not
critical.   R1 serves to provide a small purge flow around Tap A when it is
closed.   This prevents backflushed components from diffusing into the dead
section of tubing between Tap A and the T-piece.

Provided that PR2 is able to maintain its pressure when supplying flows
varying from a few ml/min to the total forward and backflush flow, there
should be no baseline disturbance when changing from forward flow to
backflush and vice versa.   The low flow provided via PR2 when in forward
flow prevents the connection to the junction T-piece from acting as a dead
leg.   The optimum setting of PR2 will be found by experience – around 3 kPa
(0.5 psi) above natural junction pressure would be a good starting point.

5.5.2   Heart-cutting (Deans 1968b)

When using columns of different characteristics in series, a pair of
components which are not resolved on the first may be separated on the
second.   If the sample is complex then other components which are separated
from this pair on the first column may co-elute with either one or both of
the pair at the outlet of the second column.   There is an obvious advantage
in allowing only that part of the chromatographic band which contains the
components of interest to enter the second column with other potentially
interfering components being diverted to vent.   This is the technique of
heart-cutting, and it is applicable to much more complex problems than the
separation of two components of a mixture.

Figure 5.18 shows the configuration for both heart-cutting and backflush to
vent.   With Tap C closed, PR1 and PR2 are adjusted as for backflush (see
5.5.1).   Tap C is opened, and R2 adjusted to offer a slightly lower
resistance than column 2.   This ensures that all the effluent from column 1
goes to vent.   A small amount of carrier gas from PR2 also goes through Tap
C to vent, purging the connection between the columns.

To cut a group of peaks from the middle of the chromatogram developed by the
first column, the sample is injected with Taps A and C open and B closed.
Tap C is closed during the period when the components of interest are
eluting from column 1.   Depending upon the nature and quantity of the
heavier components, they can be eluted forward by reopening Tap C, or
backflushed by closing Tap A and opening Tap B.

Timing of the cutting operating can be established by trial and error, or by
recording a chromatogram from column 1.   This can be done by fitting a low
dead-volume thermal conductivity detector immediately at the outlet of
column 1, or by bleeding a small amount of carrier gas from a T-piece at
this point via a length of capillary tubing to a remote detector.

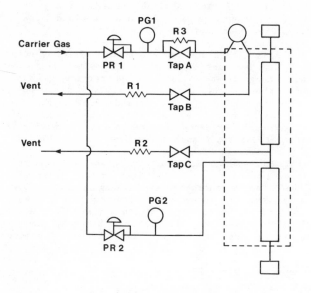

Fig. 5.18.   Heart-cutting

5.5.3   Column Selection (Deans, Huckle and Peterson 1971)

When an analysis requires that components be separated on more than one
column, pressure balancing can be used to divert groups of components onto
the appropriate column.   Carrier gas flow through the columns is necessarily continuous rather than interrupted as described in 5.4.4.

Deans and co-workers suggested a three-column scheme to deal with this and
similar problems (see Fig. 5.19).   The first column separates the sample
into three groups of components :   those amenable to separation on column 2,
those amenable to separation on column 3 and heavier components.   In their
example, quoted here to explain the operation, column 1 is Porapak T, column
2 molecular sieve and column 3 Porapak S.   Column 1 separates the sample
into a first group of hydrogen, air, carbon monoxide and methane, which are
passed to the molecular sieve, a second group of carbon dioxide and C_2
hydrocarbons, which are passed to the Porapak S, and a third group of higher
hydrocarbons and water vapour which are backflushed.

PR2 must be set to a value which is higher than the natural junction
pressure when column 1 is used with column 2 alone, or with column 3 alone
i.e. when the outlets of columns 3 and 2 are blocked off in turn.   This
ensures that the total effluent from column 1 is diverted to one of the two
secondary columns while the other is supplied with pure carrier gas.   Thus,
with Tap C in the up position, as in Fig. 5.19, all column 1 effluent goes
to column 2 : with Tap C down, it goes to column 3.   R2 is adjusted to give
a small flow to purge the connection from Tap C to the column junction which
is not taking the major flow at any time.

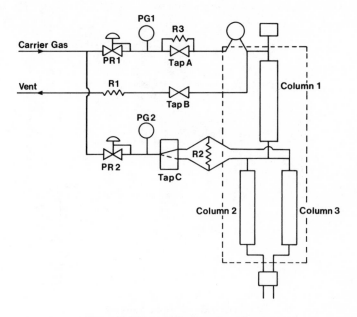

Fig. 5.19.   Column selection

Backflushing is carried out in the normal way by closing Tap A and opening Tap B.   The position of Tap C does not affect the backflushing procedure in any way.

5.5.4   Peak Storage

Heart-cutting, as described in 5.5.2 allows a group of components which are poorly resolved on the first column to be separated on a second column with different characteristics.   If a sample contains more than one such group, the task becomes complicated, particularly if there is a danger that a component from the second group which moves quickly in the second column may catch up with a slow moving component of the first group.

This problem can be avoided by repeating the analysis, with different fractions being cut onto the second column on each occasion.   Alternatively (Goode, 1977) the flow can be stopped on the first column so that groups of components are allowed onto the second column only after all the components from the previous group have eluted.   Figure 5.20 shows the layout.

Tap C, when opened, applies the same pressure to both ends of column 1, stopping the flow.   Since it does this by increasing the junction pressure, any components which are near the junction are pushed back a little way onto column 1, reducing any danger that they might diffuse across the junction. Elution of components from column 2 continues at an increased gas flowrate. When Tap C is closed, flow is restarted in column 1, and the next batch of

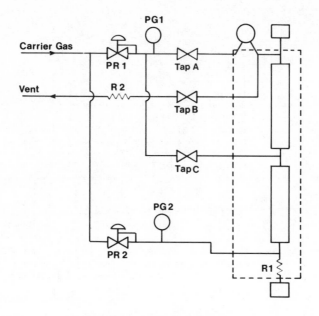

Fig. 5.20.   Peak Storage

components are allowed into column 2.   With Tap C open, heavy components
may be backflushed through Tap B in the usual manner.

If the change in flow through column 2 causes a baseline disturbance, then a
second pressure regulator, PR2 and restrictor R1 may be used.   With the
flow to the detector high, the pressure on PG2 is noted.   PR2 should be set
to a slightly higher pressure, ensuring a constant flow through R1.

5.5.5   Adjustment of Separation Characteristics

Even in cases where heart—cutting or backflushing are not required, two or
more columns with different characteristics may be necessary for separation
of all components of a mixture.   The column sizes and packings can be
selected on the basis of known retention data, but it is likely that some
final adjustment to the ratio of the columns may be necessary.   Rather than
perform physical alteration it has been shown (Deans and Scott, 1973), that
adjustment of junction pressure and hence flow rates can have the same
effect, using the layout of Fig 5.21.

The method relies upon the fact that the adjusted retention time of a
component, t'_r, varies with its hold-up time, t_m, according to the
equation

$$t'_r = t_m (K/\beta)$$

where K is the partition coefficient and β the phase ratio.   Changing t_m in one of a pair of columns but not in the other will affect its overall retention.   Where a pair of components have different retention behaviour on the two columns, changing t_m on one column will affect the ratio of their overall retentions.

Physically changing the size of one column affects t_m on that column. Alternatively, increasing or decreasing the flowrate through one column while leaving the flowrate through the other unchanged has the same effect. In fact, with the layout of Fig. 5.21 it is usual to adjust the junction pressure, simultaneously increasing t_m on one column while decreasing it on the other;   the principle is unaffected.   If the influence of column 1 should be increased to improve separation, the junction pressure should be raised.   This slows the flow in column 1, increasing t_m in that column, and increases the flow in column 2, decreasing its t_m.   Lowering the junction pressure below its "natural" value increases the influence of column 2.

Junction pressure is increased by simple adjustment of PR2 so that PG2 reads a value higher than the "natural" pressure.   It is decreased by opening the adjustable restrictor R, so that the effluent from column 1 splits between column 2 and R.   PR2 is set now to the desired value.   The setting of R is not critical.   PR2 supplies the desired junction pressure and also any extra carrier gas required if R is more wide open than strictly necessary.

This technique can be combined with backflushing and heart-cutting as described in 5.5.1 and 5.5.2.

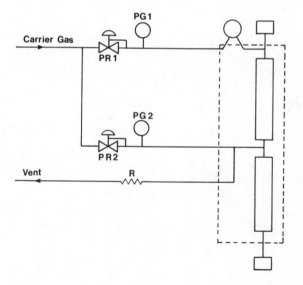

Fig. 5.21.   Adjustment of separation characteristics of two columns

5.5.6   Advantages and Disadvantages of Pressure Balancing

Little modification is needed to an existing chromatograph in order to use
pressure balancing techniques. Inside the column oven, a few fittings and
short lengths of connecting tubing are all that is required. These can
usually be chosen to be similar to the column tubing itself, and hence be
equally compatible with sample components. The small extra dead volume
introduced has a negligible effect on the separation. Extra needs outside
the column oven are pressure regulators, on-off valves and needle valves or
other restrictors. These are readily available and inexpensive.

Setting up the techniques is simple and fairly rapid even with the trial-
and-error method. Once conditions have been established, they are most
conveniently controlled by automatic timers, which ensure a high degree of
repeatability.

Since a common feature of the methods is constant flow through the second
column, or at least constant flow to the detector, the switching operations
do not cause baseline disturbances. For the same reason, there is no
practicable way of isolating a group of components in a second or subsequent
column, as in 5.4.4. Section 5.5.3 describes how groups of components are
diverted onto appropriate columns: once there, however, their elution
continues. The two secondary columns must be sized so that components do
not co-elute.

If significant amounts of high-boiling components are backflushed they may
condense in the on-off valve or the restrictor, blocking or reducing the
backflush flow. The chance is negligible if the sample is gaseous, but a
liquid sample containing dissolved gas could give trouble. The problem can
be avoided if a length of tubing loosely packed with an adsorbent (e.g.
active carbon) is inserted in the backflush line inside the column oven.

5.6   TEMPERATURE PROGRAMMING

The effect of operating a column isothermally at different temperatures has
been described in section 5.2. Whereas the procedure involves optimised
separation of different groups of components at the different temperatures,
temperature programming allows good separation of all the components of a
mixture from a single injection. Figure 5.22 shows a temperature programmed
analysis of the same mixture which was separated isothermally in Fig. 5.1.

Porous polymer beads are probably the most popular packing for this
application. Where a mixture contains light components, such as the
constituents of air, programmes commonly start at sub-ambient temperature.

Whereas the techniques described in the previous sections either require or
are well adapted to pressure control of the carrier gas, instruments for
temperature programming are almost invariably fitted with mass flow
controllers. Chapter 2 contains discussion on the relative merits of
pressure and flow control, and we only restate here that flow control does
not necessarily produce best column performance, but is needed where
detector sensitivity or baseline signal will be adversely affected by
changes in carrier gas mass flow.

Most of the procedures and precautions associated with temperature programming generally apply to gas analysis. The widespread use of the

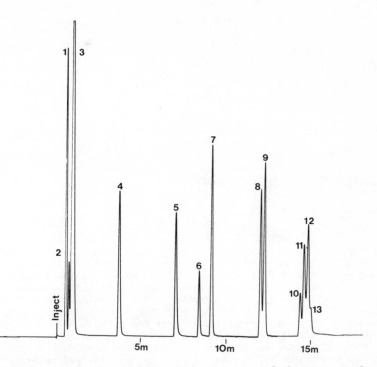

Fig. 5.22.   Temperature programmed chromatogram of the same complex
mixture used for Fig. 5.1.

thermal conductivity detector, however, means that spurious peaks due to air
components or water vapour can be more of a problem.   Unlike an FID, the
detector responds to them and they may also be sample components.   Air and
water vapour can diffuse through plastic carrier gas lines and poorly made
connections.   The carrier gas should be carefully cleaned and dried, and
only metal tubing and fittings used downstream of the purification system.

CHAPTER 6

Examples of Applications

6.1   INTRODUCTION

Few gas mixtures fall neatly into one of the classifications below.   For
example air (Section 6.2) can contain halogenated tracers (Section 6.10) in
addition to various pollutants.   Natural gas (Section 6.3) commonly contains
odorants (Section 6.7).   The purpose of the classification is to indicate
how each type of gas mixture is generally handled (Fuel gases, Refinery
gases), or to provide an overall category for gases of a particular chemical
type (sulphur-containing, nitrogen-containing) although these may be major
components of a mixture or trace components in a mixture which is itself
described under another heading.   This is justified since different parts
of an analysis require different approaches and may be needed for quite
different reasons.   Distributed natural or fuel gases usually contain
sulphur compounds as odorants, but these are ignored when the gases are
analysed for the purpose of, for example, calorific value determination.
Conversely when the odorants themselves are analysed, the analyst does not
interest himself in the major components of the mixture.

6.2   AIR COMPONENTS AND THE RARE GASES

The composition of dry air is (Weast, 1977b):-

Nitrogen	78.08%
Oxygen	20.95%
Argon	0.93%
Carbon dioxide	0.033%
Neon	18.2 ppm
Helium	5.2 "
Krypton	1.1 "
Xenon	0.09 "
Hydrogen	0.5 "
Methane	2.0 "
Nitrous oxide	0.5 "

Water vapour is also present at a level which depends on atmospheric
conditions, but is unlikely to exceed 3%.   Other components, present due to
human activity or natural phenomena, are generally called pollutants, and
are not within the scope of this section.

Air components and the rare gases are stable under virtually all circumstances that could be envisaged during sampling and analysis. The difficulty is not so much concerned with loss of these components, but their
gain, as they are the most likely contaminants in any other type of gas
sample. The ability of gases to diffuse through rubbers, plastics and
small leaks means that not only can sample components contained in vessels
or passed through tubes of these materials diffuse out, but also that air
components can diffuse in.

The thermal conductivity detector is the most popular in this field. Oxygen
eluting from the column and passing over the hot (up to $500^{\circ}C$) filaments
can oxidise their surface, causing a step baseline change. If the sample
also contains a reactive component such as hydrogen sulphide, the successively oxidised and sulphided state of the filament is demonstrated by baseline shifts in opposite directions after the elution of each type of peak.
Oxidation can be avoided by using noble metal filaments, or a thermistor
detector, the elements of which operate at much lower temperature.

The ultrasonic detector is also suitable. The detector cell does not
contain components liable to oxidation, nor does it contain high temperature
components such as filaments, and it can in fact be used with oxygen as
carrier gas.

The helium ionisation detector responds to all these components other than
helium itself, but its extreme sensitivity and limited dynamic range are
drawbacks when major components other than helium are present in a sample.

The use of molecular sieve to separate oxygen and nitrogen is probably the
best-known application in chromatographic gas analysis. Other columns can
be used, but only molecular sieves allow columns of normal length to operate
without the need for sub-ambient temperatures. Determination of argon
presents more of a problem, because on a molecular sieve column its retention time is very similar to that of oxygen, and so argon/oxygen separation demands conditions which cause nitrogen to have a long retention time.
Either way long columns or low temperatures are required. Figure 6.1 shows
a sub-ambient separation. The elution of nitrogen can be accelerated by
temperature programming after argon and oxygen have been detected, but if a
series of analyses are required this may not offer any advantage since the
total cycle time of temperature programme, cool-down and restabilisation may
be longer than the isothermal analysis time. Subambient operation of a
porous polymer column gives an opposite order of elution to that on
molecular sieve (see Fig. 6.2). This only offers an advantage if nitrogen
is a minor component. Argon is partly resolved from oxygen, but this is
visible only when the concentrations of the two gases are similar.

The problem of argon/oxygen separation can be solved by using very long
columns or sub-ambient temperatures as described above, by using argon
carrier so as to measure only the oxygen, by assuming, as Smith and Dowdell
(1973) did when analysing soil gases that the argon/nitrogen ratio remains
constant, or by converting the oxygen into another measurable species.
This last option was described by Lovelock, Charlton and Simmonds (1969) as
an application of the palladium transmodulator (see 2.6.3) : oxygen in the
sample reacts quantitatively with the hydrogen carrier gas on the surface of
the heated palladium alloy, and the resulting water is separated from argon
by a short column between the transmodulator and the detector.

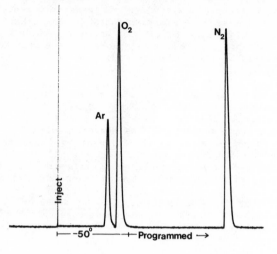

Fig. 6.1   Argon–oxygen–nitrogen separation on molecular sieve

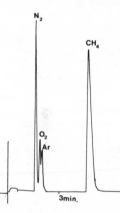

Fig. 6.2   Argon–oxygen–nitrogen separation on porous polymer

Measurement of carbon dioxide is incompatible with the use of molecular sieve, due to its extremely long retention time. Temperature programming has been described for mixtures containing percentage levels of CO_2, but the peak shape is poor, and hence the quantitative accuracy is doubtful. Porous polymer columns are widely used for CO_2 determination – it is therefore possible to analyse for N_2, O_2 and CO_2 by temperature programming on a single column, but the major component separation is not as good as on molecular sieve.   This need for different columns to be used   because of the limitations imposed by the different components in the sample is satisfied by multi-column operations as described in Chapter 5.   A porous polymer and a molecular sieve column can be used in series (5.3) or with column isolation using a valve (5.4.4) or column selection by pressure balancing (5.5.3).

The rare gases, helium, neon, krypton and xenon can be separated on 5A molecular sieve at normal temperatures.   Aubeau, LeRoy and Champeix (1965) found conditions which resolved krypton and xenon from all other components eluting from molecular sieve, and Thompson (1977) has shown separation of helium and neon, and of each from hydrogen.   Lovelock, Simmonds and Shoemake (1971) describe measurement of atmospheric helium, neon, argon and krypton using a micro-thermal conductivity detector.   The problems consist of an incompatibility in sample size between that required for the major components of air and that for the rare gases, and the interferences between oxygen and argon, and nitrogen and krypton.   This was solved by the use of the palladium transmodulator (see 2.6.3).   Nitrogen was used as the second carrier gas, thereby eliminating that interference, and oxygen was removed by reaction with the hydrogen first carrier over palladized molecular sieve. A 10ml sample allowed repeatable detection even of krypton.

Water vapour can be analysed by gas chromatography, but the various alter- native methods listed by Verdin (1973) mean that it is not the automatic choice.   Water elutes from porous polymer bead columns and is included in many example chromatograms, but the peak tails slightly, which means that quantitative accuracy must be uncertain at the levels likely to be found in an air sample.   Chromatographic water analysis is best carried out in a glass column containing poly-(ethylene glycol) coated on a P.T.F.E.   sup- port.   This column will not separate any of the other air components or rare gases.   Tactically it is often better to ignore water and to calculate results on a dry basis.

6.3   NATURAL GAS AND ASSOCIATED GAS

Natural accumulations of gas take many forms, including very pure nitrogen. In this section we consider those natural gases which may be used as fuels. Just as natural gas fields always contain some light oil (condensate), so oil fields always contain some associated gas.   Both are considered in this section, as the differences between them are not differences in components but in concentration ranges.

They contain inert gases, in particular helium, nitrogen, and carbon dioxide, and saturated hydrocarbons from methane to a carbon number which depends on the source and the treatment.   Hydrogen sulphide and other sulphur compounds may be present, either naturally or as added odorants – these are considered in section 6.7.   Water is present, usually at very low levels, and the treatment process may add small amounts of glycol and/or methanol.   Other components which may be present at trace levels are argon and ethene, but these are of specialised interest and not considered here.

Methane is usually the major component of natural gas ($\sim$90%), although from a few sources the nitrogen content may be comparable or even higher.   Other hydrocarbons are present in concentrations which, with very few exceptions, diminish as the carbon number increases.   Natural gas which has been treated for transmission and distribution contains hydrocarbons to C_9 or

C_{10}, as measurable by normal chromatographic procedures, whereas untreated gas can contain hydrocarbons to C_{16}. The composition of associated gas varies with the conditions under which it is separated from the oil, but in general any gas separated at more than a few atmospheres pressure will contain methane as the major component but significantly higher levels of C_2 and heavier hydrocarbons than a natural gas.

Apart from the sulphur compounds, which are dealt with in 6.7, natural or associated gas components are unreactive and stable. The higher hydro-carbons can be adsorbed on to rubber or plastic, and untreated gases may lose components by condensation if the sample container is stored at a lower temperature than that at which the gas was sampled. Sampling is compli-cated by retrograde behaviour, whereby heavier hydrocarbons which are stable in the gas phase at high pressure can condense as the pressure is reduced isothermally (Cooper and co-workers, 1968).

The thermal conductivity and flame ionisation detectors are the most popular. The TCD is necessary for the inert gases and has a sensitivity adequate for hydrocarbons up to C_5. The FID is necessary for the low levels of C_6 and higher hydrocarbons and has the advantage of a predictable response: it is difficult to prepare gas phase calibration mixtures containing these heavier components, and the FID allows their responses to be calculated from those of the lighter components.

ASTM D 1945-64 quotes a number of gas-liquid columns for separating lighter components of natural gas. The most popular consists of a 9 metre column packed with 28% silicone oil on Chromosorb P. This column is also spec-ified in ISO 6568, and separates N_2, CH_4, CO_2, C_2H_6, C_3H_8, $i-C_4H_{10}$, $n-C_4H_{10}$ and the pentanes. The column is somewhat cumbersome, and a similar sep-aration can be achieved on a shorter porous polymer column. A 3 metre column packed with Porapak T gives a good separation - the relatively high polarity of the material reduces the interference of the large methane peak on the small CO_2 peak.

The C_6 and higher hydrocarbons can be backflushed and measured as a group, but being present at low concentration will give a small peak if backflushed in the conventional way (section 5.4.3.1). Backflush and sequence reverse with a very short pre-column (see section 5.4.3.3) can elute the C_6+ group as a sharp peak at the beginning of the chromatogram, considerably improving its detectability. The functions of gas sampling and backflushing in this way can be performed by a single 10-port valve - Fig. 6.3 shows a typical chromatogram.

An interesting feature of this method is the use of a high carrier gas flowrate (35 ml/min for a 2mm i.d. column), which improves the nitrogen/methane separation by reducing extra-column band broadening.

The single column separation described above will not identify air contam-ination, since oxygen and nitrogen are not resolved, and, depending up their concentrations, may not separate the first few components with best quanti-tative accuracy. Improved separation can be achieved by a multi-column system, using valves for gas sampling and backflush (see section 5.4.6.1) and for column isolation (see section 5.4.4). Figure 6.4 shows a typical configuration. The column packings and lengths are chosen so that while they all operate at the same temperature, each is best adapted to the separation of a particular group of components in the mixture. Column 1 is a silicone oil column, through which permanent gases, C_1 and C_2 pass rapidly, and from which C_3, C_4 and C_5 and later C_6+ elute for measurement.

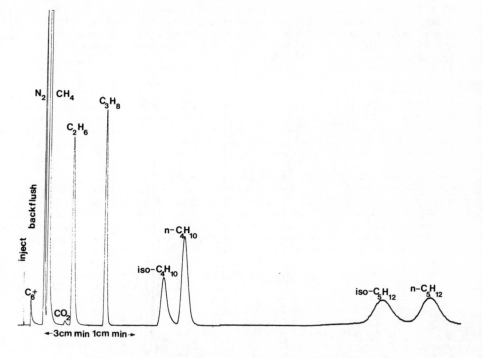

Fig. 6.3   Natural gas analysis with backflush

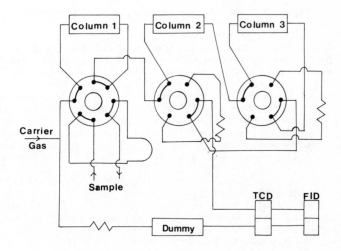

Fig. 6.4   Multi-column natural gas analyser

Column 2 contains a porous polymer, which retains CO_2 and C_2H_6 but allows O_2 (if present), N_2 and CH_4 through to column 3, which contains molecular sieve.   Each group of components passed on to columns 2 and 3 can be isolated and then eluted at a convenient time later in the run.   A typical chromatogram is shown in Fig 6.5.   Low levels of C_6+ can be better measured on a FID placed in a series with the TCD which is used for the rest of the analysis.

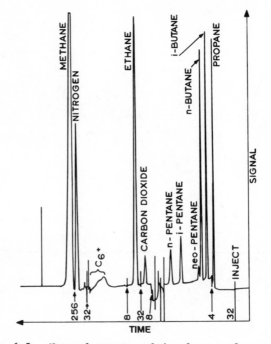

Fig. 6.5   Natural gas on multi-column analyser

A porous polymer bead column can separate each of the groups of components which are dealt with by different columns in the previous example, but at different temperatures.   Thus at -50^oC nitrogen and oxygen are separated, and at 240^oC, which is the upper temperature limit for such columns, C_8 hydrocarbons are eluted.   Stufkens and Bogaard (1975) describe such an application for analysis of distributed natural gas.   TCD and FID are used in series, and the method produces more detail of the higher hydrocarbons than either of the previous examples.   Figure 6.6 shows a chromatogram.   The method is equally well adapted to analysis of associated gas, and is described for this purpose in IP 345/80.

It was pointed out in Chapter 3 that 13X molecular sieve is capable of separating paraffins and naphthenes by carbon number, and that this could be applied to the hydrocarbons in natural gas.   Figure 6.7 shows a chromatogram.   This separation, relying on a different mechanism, provides an interesting contrast to that on a porous polymer column.   Benzene coelutes with C_9 naphthenes, but it is evident from the ratio of the other components that the C_9 naphthene contribution is very small in the case of this particular natural gas.

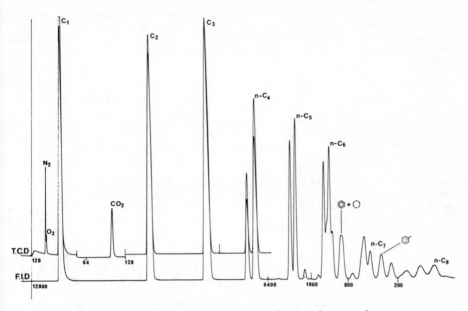

Fig 6.6   Temperature-programmed analysis of natural gas

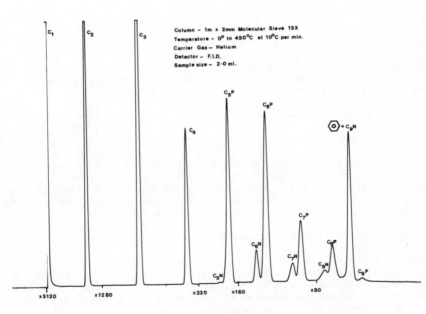

Fig. 6.7   Natural gas analysis on 13X molecular sieve

6.4 REFINERY GAS AND LIQUEFIED PETROLEUM GAS (LPG)

Oil refinery operations are extremely complex and interdependent, and so there is no particular composition which is referred to uniquely as refinery gas. The term describes a variety of gas mixtures, many of which are streams feeding other processes, and some of which are products of simpler composition, but in which the impurities can themselves be complex mixtures.

Distillation produces ethane-rich fuel gas and LPG streams of propane, butane or blends of these.   Cracking produces propene and butene streams, in addition to ethene and butadiene.   Steam reforming or partial oxidation gives synthesis gas, containing hydrogen, methane, carbon monoxide and carbon dioxide, which is considered in the next section.   A refinery gas can thus be considered to contain some or all of the saturated and unsaturated C_1-C_5 hydrocarbons.

These mixtures are stable when in contact with most materials likely to be used in analysers and sampling system.   Rubber or plastic should be minimised to avoid adsorption of C_3, C_4 and C_5 components.   LPGs can be handled as liquids under pressure (see section 2.3), in which case materials compatible with light hydrocarbon liquid are appropriate.

For detection, thermal conductivity and flame ionisation are most often used.   Most refinery gas streams contain only hydrocarbons, in which case the FID has the advantage of being flow-insensitive when valve switching is required.   The sample size when an FID is used needs to be smaller than normal (0.05 ml, or less) so that overload by a major component may be avoided.   If LPG samples are injected in the liquid phase, currently available liquid sampling valves may not allow a sample size sufficiently small to avoid possible FID overload, and a TCD would be needed.

LPGs which are straight distillation products from crude oil or condensate are easy to analyse, since saturated C_2-C_5 hydrocarbons may be separated isothermally on a long partition column or a normal-length column containing porous polymer beads (see Natural Gas, section 6.3).   The presence of unsaturated hydrocarbons complicates the situation, since on most porous polymers, C_2 and C_3 unsaturates are resolved from saturates, but butenes and pentenes overlap considerably.

Separation of the unsaturated C_4s, which can be substantial components of some samples, represents the most difficult part of the problem :   unsaturated C_5s are more numerous, but are likely only to be minor components of a gas sample.   Standard methods, such as IP 264/72 (which is identical to ASTM D 2163-77) and IP 194/74 (which is identical to ASTM D 2593-73) offer a variety of partition columns, most of which are cumbersome by today's standards, and none of which singly satisfy all the needs.   Al-Thamir, Laub & Purnell (1977) optimised conditions on an alumina column coated with a stationary phase for separation of all C_1-C_5 hydrocarbons.   The analysis takes several hours, and the full range of components includes some which are unlikely to be found in the same mixture, such as alkanes and alkynes.

In one of the rare uses of capillary columns in gas analysis (Halasz and Heine, 1962), an alumina-coated glass capillary was used for separation of C_1-C_4 alkanes and alkenes.   The separation is efficient and rapid, and its application to refinery streams was subsequently demonstrated by McTaggart, Miller and Pearce (1968).   Such a column is now commercially available, and this technique could well become much more popular.

Di Corcia and Samperi (1975) resolved most C_1-C_4 hydrocarbons on columns of graphitised carbon black modified with Carbowax or picric acid.   Saha, Jain and Dua (1978) compared the separation on different Durapaks - these are Porasils with various stationary phases bonded to the surface.   Both these types of column are of normal length and produce analyses in a reasonable time.   Carle (1979) also used coated Porasil C columns to analyse $C_1- C_5$ hydrocarbons with a C_6+ backflush using the sequence reverse method described in section 5.4.3.3.   Figure 6.8 shows a typical chromatogram of a standard mixture.   The stationary phases are Carbowax 1500, ethoxyethyl adipate and oxypropionitrile, chosen in a combination appropriate to the application.

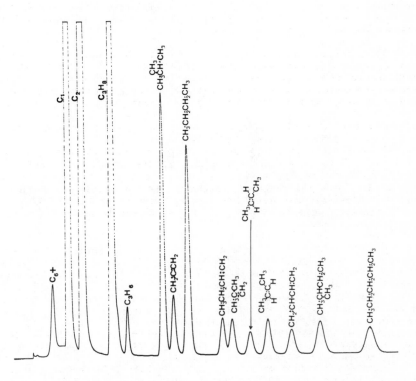

Fig. 6.8   Separation of C_1-C_5 hydrocarbons on coated Porasil

Deans and Scott (1973) also describe a three-column system for C_4 hydrocarbons.   They used squalane on Chromosorb P, Durapak (phenylisocyanate/ Porasil C) and 1,2,3 - tris(2-cyanoethoxy)propane on Chromosorb P.   They were also able to adjust the junction pressure between each column, as described in section 5.5.5. so as to "fine-tune" the separation.   When the mixtures may vary in composition, this is obviously a very attractive technique, as separations can be adjusted almost instantaneously.

6.5   SYNTHESIS GAS AND FUEL GAS

Gas mixtures for use as fuel or for chemical process feedstock can be prepared from natural gas, from LPG, from crude oil or distillate fractions, or from coal.   The production processes include carbonisation, catalytic steam reforming, hydrogenation and reaction with steam/oxygen or steam/air. Compositions vary widely, as Table 6.1 indicates.

TABLE 6.1   Compositions of Synthesis Gases and Fuel Gases

Feedstock	Coal			Petroleum fractions		
Process	Coke Oven	Air/ Steam	O_2/ Steam	High-temp reformer	Low temp reformer	GRH
H_2	54.0	19.6	28.9	64.0	15.0	8.8
O_2	0.4	–	–	–	–	–
N_2	5.6	48.0	4.4	–	–	0.1
CO	7.4	13.3	54.9	7.0	0.5	2.0
CO_2	2.0	13.3	3.4	19.0	20.0	0.2
CH_4	21.8	5.5	7.1	10.0	64.5	78.8
C_2H_6	4.7	–	0.6	–	–	10.1
C_2H_4	1.0	–	0.2	–	–	–
C_3+	2.5	–	–	–	–	–
H_2S	0.6	0.3	0.5	–	–	–

These figures show the range of components and concentrations which may be found.   The gas may be a simple 4-component mixture of H_2, CH_4, CO and CO_2, or an extremely complex one. H_2, N_2, CH_4 or CO can be the major component, depending on the process.

Hydrogen sulphide, which is present in gases made from sulphur-containing feedstock, is very reactive, and is considered in section 6.7.   The other compounds are stable, and most materials used in sampling systems are appropriate.   Any gas-making process in which steam is a reactant produces excess steam in the product, which must be removed by condensation before the gas is used.   This can affect the composition of the crude gas, as components can dissolve in the condensed water (CO_2, H_2S), or condense separately under the conditions necessary to remove the water (higher hydrocarbons).   The Appendix describes techniques for sampling the crude gas with minimum interference from the excess steam.

For this application the thermal conductivity detector is the most popular, since a major part of most fuel gases consists of inorganic components. Small but increasing levels of ethane and higher hydrocarbons in reformer gases indicate catalyst deterioration, and the flame ionisation detector is well suited to these measurements.

The components considered here are H_2, O_2, N_2, CO, CO_2, CH_4, C_2H_6 and C_2H_4. Propane and higher hydrocarbons may be present but their analysis is covered by the methods described for refinery gas and natural gas.

Simple 4-component mixtures of H_2, CH_4, CO and CO_2 can be analysed on a porous polymer column at 50°C.   This is only possible in the absence of air components, which would interfere with the CO peak.   If hydrogen accounts for most of the sample, it can be measured by difference.   If it

is desired to measure hydrogen, then use of mixed hydrogen/helium carrier
gas would be preferable to the use of argon (see section 4.7.2).   Alternatively, a h drogen transfer system can be used (see section 2.6.2).

More complex ixtures require analysis on two columns – a molecular sieve
and a porous)lymer column are the most obvious choice.   These can be used
individually,)r, better, combined by using valves as described in 5.4.4, or
by pressure b .ancing as described in 5.5.3.

Figures 6.9 ἐ d 6.10 shows chromatograms produced using these two systems.
It is worth r :ing that the elution order of CH_4 and CO is different, and so
the choice   : method should take account of any large difference in
concentration between these components.

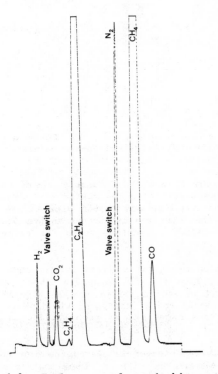

Fig. 6.9   Fuel gas – valve switching method

6.6   FLUE GAS

Flue gas analysis shows the efficiency of combustion, allows heat balance
calculations and indicates the type of atmosphere present in the furnace.
Oxygen, carbon monoxide, carbon dioxide and uncombusted fuel components
indicate efficiency, and oxides of sulphur and of nitrogen show possible
pollution problems.

AGC-D*

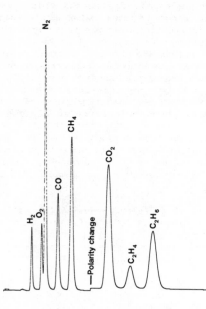

Fig. 6.10   Fuel gas – pressure balancing method

In fact, chromatography is not the automatic choice for flue gas analysis, as continuous analysers such as paramagnetic for oxygen and non-dispersive infra-red for carbon monoxide and carbon dioxide are well-proven in this area.   BS 3048 (1958) describes continuous analysers for this application, and BS 1756, Part 4 (1977) describes tests for oxides of sulphur and of nitrogen in flue gases.

Those components which might be analysed by gas chromatography are stable and unreactive.   Any flue gas will contain a lot of water vapour, and hence condensation is bound to occur, and is usually allowed for in the sampling line.   Oxides of sulphur and of nitrogen will dissolve in this condensate, making it acid, so the materials of the sampling line must be corrosion resistant.

The thermal conductivity detector is needed for many components.   Unburned hydrocarbons can be measured with a flame ionisation detector, but the levels of interest probably fall within the range of TCD.

Oxygen and carbon monoxide analysis is best performed on molecular sieve, whereas carbon dioxide requires a polymer bead column.   Either of the multi-column techniques described in the previous section would be suitable. The sample size can be adjusted as necessary to allow for different compositions.

6.7   GASEOUS SULPHUR COMPOUNDS

Hydrogen sulphide, sulphur dioxide, carbonyl sulphide, carbon disulphide and the lower mercaptans and sulphides are very important because they appear in chemical processes and gaseous emissions and because they are responsible for the odour of many materials either adventitiously or by design. Analysis is required over the range from the percentage to the sub parts per million levels.

Although chromatography has been used successfully for measuring hydrogen sulphide and sulphur dioxide, it is worth considering alternative analytical techniques. These sulphur compounds are quite chemically reactive and can easily be lost on the surfaces of equipment. Special precautions must be taken to prevent losses, particularly when analysing for traces.

Pure aluminium, which has been throughly degreased and dried, has been found by the authors as suitable material for columns and sample transfer lines for the analysis of traces of these components. Alternative materials are treated stainless steel (Pearson and Hines, 1977) and Teflon (Walker, 1978).

For the higher levels of these compounds, a thermal conductivity detector is suitable; platinum filaments are preferable if there is to be frequent exposure to hydrogen sulphide. For levels below 500 ppm, the flame photometric detector is appropriate because of its sensitivity and selectivity.

For process samples the authors have effectively measured COS and CS_2 in fuel gas on a 2m by 2mm i.d. column of 15% tricresyl phosphate on 60-80 mesh Chromosorb W at ambient temperatures after electing to measure H_2S by non chromatographic means and removing it from the sample stream by scrubbing with acidified aqueous cadmium acetate. The separation took about 10 minutes.

Kremer and Spicer (1973) used the same stationary phase but at a level of 30% on a Chromosorb P support in switchable 10 ft and 20 ft columns at 22°C. They successfully separated COS, H_2S, CH_3SH, SO_2, C_2H_5SH and CS_2 (in that order) in just over 1 hour.

The disadvantage of a liquid phase for this application is the low working temperature, though cryogenic operation makes it now practical. Higher temperatures were used by Pearson and Hines with a specially treated silica gel and a modified Porapak QS. A 1.8m column of special silica gel separated COS, H_2S, CS_2 and SO_2 (in that order) in 5 min at 60°C, whereas the same length of 80-100 mesh Porapak QS modified with 5% silicone oil at 65°C reversed the order of H_2S and COS but retained CS_2 and SO_2 longer than is convenient for measurement. Different elution orders can be used to advantage when component concentrations are widely different.

Walker (1978) used a 1 metre column of 35-60 mesh Tenax GC at 100°C and successfully separated H_2S, COS, SO_2, CH_3SH and $(CH_3)_2S$ (in that order) but does not report CS_2. Applebury and Schaer (1970) measured H_2S, SO_2 and CH_3SH in the stack gas of a Kraft pulpmill on a "continuous" basis. They used a 6ft, ½inch o.d. column packed with Porapak Q at 90°C and needed a 40 ml sample because of the nature of their detector (a home-made colorimeter) and some losses in the system (especially SO_2). Presumably the column could be used with a flame photometric detector to overcome the sample size problem.

Goode (1970) measured the indigenous sulphur compounds in natural gases.

He reports using tricresyl phosphate on celite and squalane on celite to separate a range of volatile sulphur compounds.   Mercaptans were oxidised to disulphides in order to simplify the identification of both sulphides and mercaptans.   Gibbons and Goode (1968) measured the sulphur compounds added as odorants to natural gas.   Using a flame photometric detector, the odorants must be separated from each other and also from the major component hydrocarbons in the gas so as to avoid quenching.   They first described the use of trixylenyl phosphate at ambient temperature, but subsequently preferred OV-17 (1m x 2 mm i.d., 15% w/w on 80-100 mesh celite) at 40°C, giving an elution order of CH_3SH, C_2H_5SH, $(CH_3)_3CSH$, $CH_3SC_2H_5$ and $(C_2H_5)_2S$. For the commonly used odorant thiophane (tetrahydrothiophene, THT) a very short (10 cm) column was sufficient.   (Private communication).

6.8   GASEOUS NITROGEN COMPOUNDS

6.8.1   The Separation of Nitrogen and its Oxides

Many attempts have been made to produce a scheme to separate N_2, N_2O, NO and NO_2. N_2 and N_2O are unreactive and are easy to deal with.   NO is prone to give tailing peaks and readily reacts with O_2 to give NO_2. NO_2 is reactive and is a problem to handle and separate.   Porapak Q has the right partitioning characteristics for the separation but NO_2 can react with it (Trowell, 1971) and so it cannot be recommended.

Either a two column system or temperature programming is necessary in the absence of a suitable single isothermal column.   Lawson and McAdie (1970) examined the combination of either silica gel or molecular sieve 5A with SF96 on Fluoropak-80 for separating NO and NO_2 from air.   They concluded that chemical reaction prevented the direct measurement of NO and recommended that it be positively oxidised to NO_2 and the increased level of NO_2 measured using a 20ft by 1/8 in column of 10% SF96 on Fluoropak-80 at 25°C.

Dietz (1968), after a very elaborate column preparation, used molecular sieve 5A (6ft by ¼" o.d.) to separate N_2, N_2O and NO.   Clay and Lynn (1975) further elaborated the column treatment to improve the peak shape.   Smith and Chalk (1979) used 0.6 m by 5mm i.d. of 100-120 mesh molecular sieve 5A to separate N_2, NO and N_2O, in that order, by temperature programming from 35 to 250°C.   The presence of oxygen will prevent the measurement of NO.

Other, non-chromatographic, methods are available for measuring NO and NO_2, such as the chemiluminescence method (Allen, Billingsley and Shaw, 1974) and the prefered tactic would be to use one of those methods and to restrict chromatography to the measurement of N_2 and N_2O.

Of these oxides, only N_2O can conveniently be measured by the use of an electron capture detector (see section 2.5.3).

6.8.2   The Separation of Ammonia and the Methylamines

Ammonia and the methylamines may be separated on many column materials but tailing due to hydrogen bonding reduces their effectiveness.   However, Applied Science Europe BV (1979) report the use of the proprietary packing material Pennwalt 231.   With a column of 20ft by 1/8in o.d., ammonia, methylamine, dimethylamine and trimethylamine can be completely resolved in 14 min by temperature programming from 50 to 100°C.

Chromosorb 103 is specifically designed to separate amines and 3ft by 4 mm
i.d. of 60–80 mesh Chromosorb 103 at 140°C will separate ammonia from
methylamine in less than half a minute.

6.8.3   The Separation of Cyanogen Compounds

Isbell (1963) successfully separated cyanogen, cyanogen chloride and
hydrogen cyanide from each other and from a number of other permanent gases
including chlorine.   He used an 8ft x 5 mm i.d. column of 25% triacetin on
30–60 mesh Chromosorb P at 75°C.   The elution order was   (N_2O , CO, CO_2 ,
NO, CH_4 all unresolved), Cl_2, $(CN)_2$, CNCl and then HCN.

6.9   GASEOUS HALOGEN COMPOUNDS

Inorganic and organic halogen compounds are so different in handling
problems and chromatographic behaviour that they will be treated quite
separately.   Organic halogen compounds also have an important role as
tracers and this will be covered in Section 6.10.

Ordinary materials of construction of chromatographs are not suitable for
handling such reactive gases as HF and Cl_2 and special materials must be
carefully selected.   Nickel, monel metal, selected types of stainless steel
and PTFE must be the only materials contacted by the sample.   Even then
some losses can be expected and it may be necessary to condition the
equipment before obtaining meaningful results.   PTFE is the favoured tube
material for the columns (Ellis and Iveson, 1958).   Spears and Hackerman
(1968) report achieving separation with empty monel columns.

Of the general detectors, described in section 2.5, the thermal conductivity
detector has been widely used for inorganic halogen gases, though the block
and, more particularly, the filament materials must be carefully chosen.
Both nickel (Ellis and Iveson, 1958) and teflon-coated tungsten filaments
(Lysyl and Newton, 1963) have been used successfully.

With such reactive gases specialist detector systems such as the electrolytic conductivity (Coulson, 1966) and the thermionic detector (Karmen,
1964) are needed. These and others are well described by David (1974).

6.9.1   Organic Halogen Compounds

The gaseous organic halogen compounds are substituted methane, ethane and
ethylene and the less common fluorine substituents of propane (which will be
ignored here).   They include some of the "halocarbon" family of refrigerants
and the environmentally sensitive vinyl chloride (monomer).

Ratcliffe and Targett (1969) describe the analysis of dichlorofluoromethane
for impurities which include many of the gaseous organic halogen compounds
and their separation system should be adaptable for most separations of
these compounds at higher levels.   They used a 4m by 3mm o.d. column of 6%
silicone oil (Type MS550) on 80–100 mesh Chromosorb W, modified by the
addition of 6% Bentone 34, at 30°C.   The relative retention times are
given in Table 6.2.

TABLE 6.1　Retention Times of some Organic Halogen Compounds
Relative to Dichlorofluoromethane.

Formula	Trade Name	Relative retention time
CF_4	Halocarbon 14	0.51
CHF_3	Fluoroform	0.52
$CClF_3$	Halocarbon 13	0.52
$CH_2= CF_2$	Vinylidene fluoride	0.54
CCl_2F_2	Halocarbon 12	0.58
$CF_2= CF_2$	-	0.58
$CClF_2CClF_2$	-	0.63
$CHClF_2$	Halocarbon 22	0.64
CH_2F_2	-	0.69
$CFCl = CFCl$	-	0.83
CCl_3F	Halocarbon 11	0.94
$CHCl_2F$	Halocarbon 21	1.00
CH_2ClF	-	1.24
CH_2Cl_2*	-	3.10
$CHCl_3$*	Chloroform	3.30
CCl_4*	-	3.62

*Liquids at room temperature.

Polymer beads are also effective in separating this group of compounds.
For example Applied Science Europe BV (1979) report that Halocarbons 12
(CCl_2F_2), 114 ($C_2Cl_4F_2$), 11 (CCl_3F), and 113 ($C_2Cl_3F_3$) can be separated,
with that order of elution on a 6ft x 2mm i.d. column of 80-100 mesh Porapak
Q at 150°C.

There is often a need to measure vinyl chloride in air; Hendifar and Tirgan
(1978) achieved a reasonable separation of vinyl chloride from other air
pollutants with a 3ft by 1/8 in o.d. column of 80-100 mesh Porapak QS at
100°C.　　However to confirm that there was no interference from a
coeluting component, they also hydrogenated the vinyl chloride to ethyl
chloride by mixing the sample with hydrogen and passing it over a palladium
catalyst on its way to the Porapak QS column.　　This yielded a peak eluting
after other pollutants.

Goldan and others (1980) found that 3.4m by 1/8 in o.d. of 80-100 mesh
Porasil C gives excellent sensitivity to vinyl chloride when used with a
chemically sensitised electron capture detector.　　This column, however,
does not resolve vinyl chloride from the lighter halocarbons and when these
are present a column of 80% Porapak S and 20% Porapak T (80-100 mesh, 1.5 m
by 1/8 o.d.) at 80°C is necessary as recommended by Krockenberger and
others (1979).

6.9.2   Inorganic Halogen Compounds

For separating inorganic halogen compounds, chemical reactivity limits the choice of stationary phase. Halocarbon oils are the most popular choice and some published retention times on these and other stationary phases are given in Table 6.3.

TABLE 6.3   Some Selected Retention Times of Inorganic Halogen Compounds.

	A	B	C	D	E
Liquid	Halocarbon Oil 11-14	Halocarbon Oil 13-21	Halocarbon Oil 13-21	Kel F 40	-
Support	Chromosorb W	Kel F powder	Kel F powder	Kel F powder	Silica Gel
Temperature	26°C	26°C	ambient	48°C	56.5°C
Relative retention times (Reference underlined)					
Cl_2	1.20	4.00	-	1.00	2.14
F_2	-	1.24*	0.93*	-	-
ClF	-	-	-	0.70	-
ClF_3	-	-	-	1.40	-
Br_2	-	-	-	2.28	-
BrF_5	-	-	-	3.05	-
HCl	-	-	-	-	1.00
HF	-	0.56*	1.47*	0.85	-
SF_6	1.00	1.00	-	-	-
NF_3	0.63	0.37	1.00	1.00	-
$COCl_2$	-	-	-	-	0.74
SiF_4	0.61	0.22	-	-	-

A+B after after Lysyl and Newton (1963)
C   after Spears and Hackerman (1968)
D   after Ellis and Iveson (1958)
E   after Fish and others (1968)

*The inconsistency of order of elution from the same column packing may be due to differing interaction with tube materials etc.

6.10   TRACERS

A tracer gas is any component added to a large mass of gas to determine its direction of movement or its flowrate.   Any gas may be used as a tracer, provided that it is not normally present in the system to which it is being added, and that it can be detected without interference.   The larger the scale of operations, the more desirable it is that the chosen tracer should be detectable at extremely low levels.   It is, of course, axiomatic that the tracer should be stable under all conditions likely to be encountered in the system under test or in the analytical device used to measure it.   These considerations mean that sulphur hexafluoride and particular halocarbons have become very popular as tracers, and the electron capture detector as the measuring device.

Chromatography is not necessary if the tracer is the only electron-capturing species present.   Even in the presence of a steady concentration of a mildly electron-capturing material, such as oxygen in air, the rate of sample entry into the detector can be controlled so as not to reduce the standing current below an acceptable value, and the presence of tracer seen as a further reduction in this standing current.   For best accuracy and greatest sensitivity, however, separation of the tracer from oxygen and from other electron-capturing components is needed.

SF_6 and BCF (bromochlorodifluoromethane) can be detected coulometrically by electron capture.   Halocarbon 12 (dichlorodifluoromethane) can be detected with lower, but still spectacular, sensitivity.   Other intensely electron-capturing materials which may be encountered are volatile liquids such as halocarbon 11 (trichlorofluoromethane) and carbon tetrachloride. SF_6 and oxygen are easily separated on alumina, porous polymer beads or graphitised carbon black (Zoccolillo and Liberti, 1975).   Simmonds and co-workers (1972) used the exclusion effect possible on 5A molecular sieve to elute SF_6 before oxygen, and hence reduce the detection limit.   Halocarbons are conveniently separated from each other and from air on a gas-liquid column such as squalane or silicone oil: oxygen and sulphur hexafluoride are not separated on such a column.

Meteorological work accounts for many applications, with SF_6 used to trace air or pollutant movements.   A technique for pre-concentrating the tracer has been described (Clemons, Coleman and Saltzman, 1968) which allows detection of one part SF_6 in 10^{14}.   This allows the tracer to be measured up to 75 miles from its point of release into the atmosphere.   Lovelock (1971) has used a chromatograph with a coulometric ECD to measure atmospheric background levels of SF_6 and halocarbon 11, and has correlated the concentration of these exclusively man-made components with air movements over Europe.

More recently, the ECD has been used again for continuous measurement of SF_6 (Simmonds, Lovelock and Lovelock, 1976).   Atmospheric oxygen is removed by catalytic reaction with hydrogen, excess hydrogen diffuses through the palladium alloy vessel containing the catalyst, and the water is removed by condensation.   The residual N_2, CO_2 and SF_6 pass to a dual ECD, the first part acting as a solute switch and the second measuring the resulting modulated signal.   Other halocarbons have been shown to be destroyed over the catalyst at several hundred times their expected atmospheric levels. This continuous measurement is well suited to an aircraft-borne analyser for assessment of plume dispersion from tall stacks (Lund Thomsen and Lovelock, 1976).

The authors have had considerable experience in adding tracers to natural gas networks, for leakage location and flow measurement. For leakage work, it is convenient to use an instrument which acts as a continuous sampler, but can be easily converted to the chromatographic mode for confirmation of an identification. Since the continuous mode involves a constant oxygen background when sampling air, enough tracer must be added to the gas to overcome what would otherwise be a negative signal due to oxygen deficiency.

Time-of-flight flow measurement is made by adding a slug of tracer and timing its arrival downstream. We have tried this over a 40-mile length of transmission main, with the detector being used in the continuous mode. This does not involve chromatography, except that the tracer could be seen to be an unadsorbed component in a large diameter open tubular column. Symmetrical gaussian peaks are obtained, and the measured dispersion of the peaks agrees well with the calculated value.

6.11   ISOTOPES

Separation of isotopes requires very high column efficiencies, and is one of the few areas in gas analysis where capillary columns have been used. Reported separations concern stable isotopes of non-reactive gases, and so there are no problems of adsorption associated with this application.

An etched glass capillary column, 47 metres long, has been used at $-196^{\circ}C$ for separation of isotopic methanes (Bruner, Cartoni and Possanzini, 1969). The carrier gas was a mixture of helium and nitrogen, with the nitrogen being claimed to act as a deactivating agent. Deuterated and tritiated methanes were well separated, but only a small degree of separation was possible between C H , C H and C H . The naturally occurring level of CH D in methane was measurable. An 82 metre etched glass capillary was used at $-254^{\circ}C$ for separation of Ne   and Ne   (Purer, Kaplan and Smith, 1969). In this case, the carrier was helium containing 5% hydrogen as deactivating agent.

High-efficiency packed columns have been described by Bruner and DiCorcia (1969) for separation of nitrogen isotopes. A 60 metre column packed with 0.1% squalane on graphitised carbon black generated 45000 plates at $-196^{\circ}C$. Carbon monoxide was added to the hydrogen carrier gas to deactivate the surface, and the separation took just over 8 hours. The same packing has been used (DiCorcia and Bruner, 1970) for deuteromethanes at $-78^{\circ}C$. A 15 metre column separates CH   and CD , whereas 120 metres is necesary for CH /CH D separation.

6.12   MISCELLANEOUS GASES

Ambient levels of phosphine in air have been analysed using a Chromosorb 102 column at ambient temperature with a flame photometric detector (Bean and White, 1977) or a long squalane on Chromosorb P column at $50^{\circ}C$ with a thermionic nitrogen/phosphorus detector (Vinsjansen and Thrane, 1978). A recent instrument (Airco, 1980) uses a photoionisation detector to monitor continuously for arsine and phosphine. It converts rapidly to a chromatographic mode to confirm identification of either or both of these gases.

Silanes, boranes and germanes, while difficult to handle and very toxic, do not present much difficulty chromatographically in terms of their separation from each other.   Either porous polymer or boiling point columns can be used.   The following are typical retention times for a 30 foot column containing 30% hexadecane on Gas-Chrom RZ at ambient temperature.

Silane	2.32 minutes
Diborane	2.58
Germane	3.34
Phosphine	3.48
Arsine	6.02
Hydrogen sulphide	6.42

Chlorosilanes can be measured in silane using a 10 foot column of squalene on Chromosorb W at 50°C (Gow-Mac application note GM-123S).   It is suggested that the use of a gas density balance eliminates the need for calibration gas mixtures.

Ethylene oxide in gas mixtures is widely used for sterilising, and is conveniently separated from other gases on Porapak Q at 130°C (Gow-Mac application note GM-117S).

CHAPTER 7

Gases in Liquids

7.1 INTRODUCTION

Many liquid samples are presented for analysis for components which are normally considered as gases, and consequently are appropriately treated for analytical purposes using the apparatus and techniques described earlier. The sample handling methods however are quite different from those used for gases.

Gases may be present in the liquid phase for different reasons, and are conveniently classified accordingly, since each type of liquefied gas has its particular constraints on handling and analysis. The types are:-

- gases liquefied by pressure at ambient temperature

- gases liquefied by temperature at ambient pressure, and

- gases in physical solution in liquids which are themselves stable at ambient temperature.

The first two categories are assumed to give no liquid or solid residue when brought to ambient conditions. Gases in solution are usually minor components, and the analytical technique must take account of the behaviour of the liquid major component.

7.2 GASES LIQUEFIED BY PRESSURE

The majority of items in any compressed gas supplier's catalogue are materials whose critical temperatures are above ambient, and so are provided as liquefied gases under pressure in cylinders. Chlorine, sulphur dioxide, ethane and nitrous oxide are typical examples. The pressure in the cylinder is the vapour pressure of the contents at ambient temperature and remains constant thoughout the life time of the cylinder while liquefied gas is still present. The cylinder outlet may be connected to either the gas or the liquid phase. Except that impurities in the gas may be distributed preferentially into one phase or the other, it is generally true that there is little difference between the gas and liquid phases, and so the contents may be sampled and analysed as a gas.

Other gases liquefied by pressure are mixtures (liquefied petroleum gas, LPG, or Freons) in which the composition of the two phases is different. As provided, the liquid phase represents the greater part of the cylinder contents by mass, and so analysis of the liquid is more important than that of the gas. A sample of the liquid can be totally vaporised and injected into the chromatograph as a gas, or injected directly as a liquid.

7.2.1   Vaporisation

Miniature sample bombs are available, made of nylon with metal valves, with capacities of up to 5 ml, and which will stand pressures up to 6.7 MPa (1000 psi).   When filling sample containers with any liquid, it is a safety requirement that some vapour space (ullage) be left above the liquid so that the container will not be damaged by hydrostatic pressure changes.   On the other hand, a true liquid-phase sample can only be guaranteed by filling the container completely with the liquid.   Nylon has the advantages of slight flexibility (by comparison with stainless steel) so that damage to the container is much less likely, and of translucency, so that the absence of vapour bubbles can be confirmed.   Nonetheless, it is good practice to minimise the time during which the sample is confined in the liquid phase.

The sample container is mounted vertically, then filled by connecting the lower valve to the sample source, opening this lower valve and the sample source valve fully, and controlling the flow through the container by means of the upper (outlet) valve.   When the container has been thoroughly purged, the sample is isolated and the container connected via its lower valve to a gas sampling bulb as shown in Fig. 7.1.

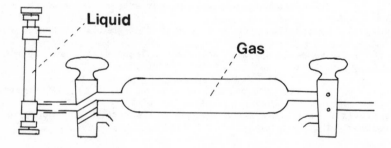

Fig. 7.1.   Transfer of liquefied gas to gas sampling bulb

The bulb and connecting tubing are evacuated up to the sample container valve, isolated from the vacuum, and the valve then opened to flash the liquid contents into the bulb.   The relative sizes of sample container and bulb should be chosen to allow for the 200- to 300- fold expansion which occurs on vaporisation.   It is obviously more convenient if the final pressure in the bulb is slightly above atmospheric, but this will depend on the composition of the sample.   If significant amounts of condensible components are present (such as pentanes in LPG), the final pressure in the bulb must be arranged to be low enough to ensure that they are not lost by condensation on the walls of the bulb.

7.2.2   Liquid Injection

Liquid sampling valves, capable of injecting samples in the microlitre range
and of withstanding the pressures involved, are available for and easily
fitted to gas chromatographs. The flow of liquefied gas through such a
valve must be controlled by restricting the outlet so as to maintain the
sample in the valve in the liquid state. Figure 2.9 in Chapter 2 shows an
example. Vaporisation of the sample after the valve is switched to the
inject position is virtually instantaneous, since the mass injected is
small. Any droplets of heavier components left in the valve will be
evaporated into the carrier gas now flowing through the sample volume in a
few seconds. This slightly slower injection of the heavier components is
not a problem since the injection band width is still only a fraction of the
final peak width. The valve should be kept at or below the temperature of
the sample, as the vapour pressure of the sample increases with temperature,
and it becomes impossible to ensure that the valve is filled with liquid
unless an external overpressure is applied to the sample. Injection of a
liquid sample is attractive since it eliminates the extra manipulation stage
of vaporising the liquid to gas.

7.2.3   Analysis

The analytical method used is not affected by whether the sample is injected
as gas or liquid, assuming that the gas or liquid sampling valve is appropriately connected to the chromatographic column.   Chapter 6 describes the
analytical procedures.

7.3   GASES LIQUEFIED BY TEMPERATURE

Gases whose critical temperatures are below ambient cannot be liquefied by
pressure alone, and so for the purposes of bulk transport and storage must
be maintained in the liquid state by low temperature.   Because of the
relative inconvenience of handling gases under these conditions, only the
permanent gases are treated in this way.

Most cryogenic gases are single substances, with varying amounts of impurities. The method of storage and the constant boil-off means that
impurities with lower boiling-points are unlikely, and that the levels of
impurities in the liquid and the gas are likely to be very different.   In
the case of cryogenic mixtures, such as liquefied natural gas, the differences between the phases is very significant.

Although sampling valves exist which can cope with temperatures as low as
that of liquid nitrogen (-196°C), we know of no example of injection in
the liquid state. Vaporisation and injection as a gas is normal.

7.3.1   Sample Handling

The principle of isolation of a liquid sample followed by complete vaporisation can be used, as for pressure-liquefied gases.   The difference in
procedure concerns the danger of isolating a cryogenic liquid for any
prolonged time.   The general design of sampler has a smaller vessel
contained within a larger one.   The ratio of their sizes is such that the
sample contained in the smaller vessel as a liquid will evaporate to fill
the larger one as a gas.   (The likelihood of condensible components being

present is small, and so the sizes are chosen such that the final gas sample is pressurised in the larger vessel).

The sample enters the smaller vessel, overflows into the larger, and is then vented. As the sampler cools, the inner vessel will become filled with liquid. At this point the sample flow is stopped, any excess liquid is rapidly drained from the outer vessel, the drain valve closed and the liquid allowed to evaporate. The disadvantages of the method, by comparison with that in 7.2.1, are that the necessary speed of the operations may not allow all the liquid to have drained from the outer vessel, and that the evaporation and subsequent homogenisation of the gas may be slow.

Another method, suitable for permanent sampling points, is to allow the cryogenic liquid to flow through a length of capillary tubing into a larger diameter tube in which complete evaporation is ensured. The dimensions of the capillary can control the flowrate, but it is important that the entire length of the capillary should be cold, which can be achieved, for example, by mounting it in a metal block which is in contact with the cryogenic liquid. If there were a temperature gradient in addition to a pressure drop, intermittent boiling of the sampled gas within the capillary could cause vapour to be blown back against the sample flow, while leaving a film of heavier components on the tubing wall. The overall composition of the stream leaving the capillary may then not be representative of the sample source.

7.3.2   Analysis

As the sample is fully vaporised before analysis, no unexpected complications arise. Mixtures such as liquefied natural gas are analysed more simply than their starting products, since the liquefaction process causes many of the heavier components to be removed.

7.4   DISSOLVED GASES

The chromatographer, with knowledge of gas–liquid phase equilibrium, is in a good position to appreciate the behaviour of dissolved gases. Any liquid in contact with a gas contains some of the gas in solution, the amount being dictated by the coefficient of solubility and the partial pressure of the gas. The solubility of gases decreases with increasing temperature, which is the same as saying that the vapour pressure of a dissolved gas above its solution increases with temperature. Thus any liquid open to the air contains dissolved oxygen and nitrogen. These components (of which oxygen is more significant) can be removed by boiling or by bubbling a different gas through the liquid. It is important to recognise that the latter procedure does not physically displace the previously dissolved components, but works because their partial pressure above the solution is reduced. The advantage of bubbling a gas through the liquid rather than passing it over the surface is that equilibrium is reached more quickly due to the agitation of the liquid. When dissolved oxygen, for example, is removed by boiling, the liquid must subsequently be allowed to cool under an oxygen-free atmosphere, so as to prevent oxygen from being slowly redissolved.

Liquids may contain gases dissolved at atmospheric pressure, in which case they can be handled by conventional means (e.g. syringe for sample injection), or at high pressure, in which case they can not. Crude oils offer a good example of the range of pressures which may be encountered.

The pressure in the reservoir varies with depth, but will be in the region of 34 MPa (5000 psi) for a well of 10,000 feet. Available pressure at the well-head depends upon the density of fluid in the riser and the flowrate, but may be as high as 6.7 MPa (1000 psi). Depending upon the amount of gas dissolved in the oil, its bubble-point (the pressure at which bubbles of gas start to separate from the fluid) may be up to 24 MPa (3500 psi), which means that gas will start to separate in the riser, and that the only way to sample single-phase reservoir fluid is by means of a "bottom-hole" sampler which is lowered down to reservoir depth.

At the well-head the oil is separated from the gas by a series of pressure-reduction stages, finally to emerge as a product which, while still containing dissolved gas components, is stable at atmospheric pressure. The gas released from the oil at each stage of pressure-reduction will vary in quantity and composition according to the gas/liquid equilibrium at each stage.

Another difference between types of sample, which dictates the analytical approach, concerns the nature of the liquid in which the gases are dissolved. This may be compatible with the chromatographic system to be used for the separation of the gases, or not. An example of the former would be hydrocarbon gases dissolved in a low-boiling solvent, such as acetone or methanol, where all the components could be separated on a porous polymer bead column. An example of the latter would be oxygen dissolved in any liquid. The column required for unequivocal separation of oxygen, such as molecular sieve, will rarely be appropriate for the liquid.

7.4.1   Handling of Stabilised Samples

Injection of the liquid sample by microsyringe, a technique not described in this book but familiar to most chromatographers, is appropriate in most cases. If the sample contains a lot of gas, it may be difficult to draw the liquid into the microsyringe as the small pressure reduction caused by withdrawal of the plunger may cause bubbles of gas to separate. This can be overcome by chilling the syringe, or by pressurising the sample with a gas other than one of the components to be analysed. Figure 7.2 shows the system.

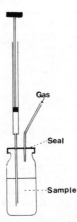

Fig. 7.2.   Sampling liquid saturated with gas

Analysis for dissolved oxygen requires special precautions, since the ubiquitous atmospheric gases can easily contaminate the sample. A modified syringe for sample transfer is shown in Fig. 7.3 (B.A. Cleaver, Private communication). The valves allow the syringe and connecting tubing to be flushed through with excess sample, which can then be isolated. The syringe contents are then flushed through a liquid sampling valve, which injects an aliquot into the chromatograph.

Fig. 7.3. Syringe for sampling liquid for measuring dissolved oxygen

7.4.2   Handling of High-Pressure Samples

Sample integrity can most reliably be preserved by maintaining the liquid under pressure so as to avoid gas separation. Safety, however, is served by allowing an ullage space so that thermal expansion of the sample does not cause a dangerous pressure increase. This means that samples are frequently taken into high-pressure sampling vessels, and then a small amount of liquid is run off, creating an ullage space. Before analysis the sample must be recombined by pressurising with an immiscible fluid (e.g. mercury). A new type of constant pressure cylinder which overcomes this difficulty is shown schematically in Fig. 7.4, in which a moving piston separates the sample from the ullage space. The gas side of the piston is pre-pressurised before the sample is taken - any appropriate gas can be used as there is no physical contact between this gas and the liquid sample.

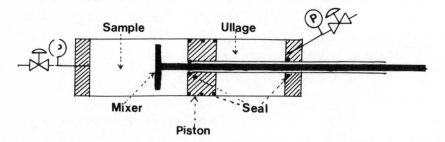

Fig. 7.4   Welker pressurised liquid sample vessel

The high pressure sample can be separated before analysis into stable gas and liquid phases, or, in those cases where the liquid is relatively low boiling, injected via a liquid sampling valve as described in 7.2.2 for gases liquefied by pressure. A schematic drawing of an apparatus for separation of high-pressure crude oils or condensates is shown in Fig 7.5 (B. Penney, and J.B. Oleksin, Private communication).

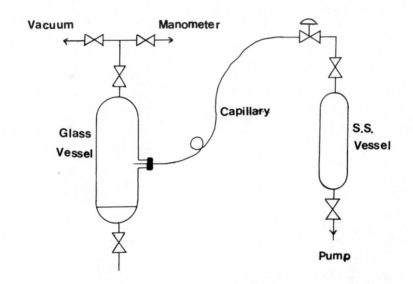

Fig. 7.5.   Apparatus for separating gas from a pressurised liquid

The sample is bled slowly into a previously evacuated bulb, where separation of the phases occurs, until the pressure reaches atmospheric.   The two phases are allowed to equilibrate, preferably overnight, and then each analysed.   Both gas and liquid must be analysed, since the amount of gas left in the liquid will be significant.   The results from the two analyses are summed, with the knowledge of the mass of oil (by weighing the bulb before and after), the volume of gas and its density (from analysis).

7.4.3   Analysis – Gas Only

In many cases it is important that the liquid should not contact the column used for separation of the gas components, because it would cause an extremely long analysis time, or interfere with subsequent analyses, or even degrade the column packing used.   This is a classic application for back-flushing to vent, as described in sections 5.4.3.2 and 5.5.1.   For dissolved oxygen, the precolumn, which may for example provide a simple boiling-point separation, need only separate the dissolved gases from the liquid, which is then backflushed while the gases are being separated on the analytical column, which is most likely to be molecular sieve.

Another application, described in IP344/80, concerns measurements of light hydrocarbons in stabilised crude oil.   Both pre-column and analytical column contain Porapak Q at 150°C, components up to C_6 being allowed forward and the remainder backflushed.   (Obviously, most of the heavier components in the oil have no mobility on Porapak Q at the temperature used and so are not removed by backflushing.   They are, however, prevented from entering the analytical column.   Replacement of the pre-column is advisable after a

certain number of analyses).   Figure 7.6 shows a typical chromatogram:
2,2-dimethylbutane is added as an internal standard.

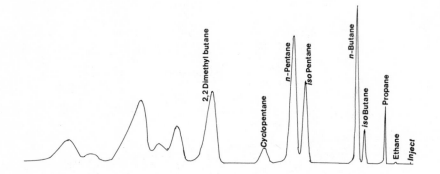

Fig. 7.6   Chromatogram of dissolved gases in a stabilised crude oil

7.4.4   Analysis - Gas and Liquid

There are obviously many applications for which the liquid and the gases
dissolved in it need to be analysed.   Analysis of liquids is outside the
scope of this book, and we mention it because of those instances where the
liquid and the dissolved gases have similar chromatographic behaviour, or
where analysis of the gases imposes chromatographic conditions such that the
liquid is more conveniently eluted through the system than removed by
backflushing.

Methanol or acetone containing dissolved C_2-C_4 hydrocarbons can be analysed,
for example, on porous polymer beads, the choice of column being dictated by
potential interferences between the gas and liquid components.   Dilution of
lubricating oil by dissolution of hydrocarbon gases is another example.
The heavier hydrocarbons which are minor components of the gas become
significant components in the liquid due to their higher solubility, and so
analysis of the "dissolved gases" needs a temperature programme to deal with
C_1 to C_9 hydrocarbons.   This being so, continuation of the temperature
programme to clear the oil from the column may be preferred to other methods
of removing the oil from the system.   Figure 7.7 is a chromatogram of such a
sample, using a 2m x 2 mm i.d. column of 5% OV-101 on 80-100 mesh Chromosorb
W, temperature programmed from -50 to 350°C.   The oil itself, which
covers the C_{20}-C_{40} range is fully eluted during this programme, leaving the
column clear for the next analysis.   It is interesting to compare Figs. 7.7
and 7.6.   The temperature programming method allows measurement of groups
of hydrocarbons up to C9, but with less good resolution of the lighter
hydrocarbons.   Backflushing, even with the limitations described, is more
suitable for crude oil, as temperature programming will not remove all the
heavy end of a crude from a column.

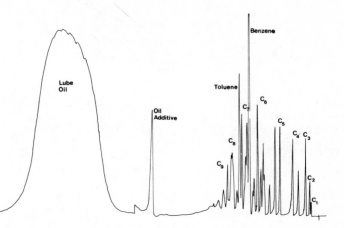

Fig. 7.7.   Chromatogram of hydrocarbon gases dissolved in a lubricating oil

7.5   HEADSPACE ANALYSIS

The problems associated with the presence of the liquid in the chromatographic system are avoided by the technique of headspace analysis, whereby
the vapour above the sample, which is rich in volatile components, is
injected under controlled conditions.   The subject has been recently
described by Hachenberg and Schmidt (1977) and is not considered further
here.

A technique of discontinuous gas extraction has been described by Kolb and
Pospisil (1977).   Volatile components in solids are measured by allowing
the sample to equilibrate with carrier gas in a by-passed injection port,
and then injecting the slug of vaporised components.   Quantitative information is gained by repeating the procedure exhaustively, or by calculation
from the first few values.

CHAPTER 8

Quantification

8.1 INTRODUCTION

Chromatograms are interpreted in two stages, the first being qualitative and the second quantitative. The qualitative stage is relatively simple in that it deals only with the location of a peak in the chromatogram, referring it to the locations of peaks due to identified components in the same and other chromatograms. The use of relative retention data, the technique of "spiking" samples before analysis, and identification through the use of selective detectors are techniques so well established as not to require further elaboration. Quantitative measurement, however, is a more complex affair - mistakes in peak identification are rare, but those in quantitation are much more widespread.

Accurate quantitation requires that the peaks are sufficiently well separated so that they do not interfere with each other, that the detector signal is a predictable function of the amount of component, and that this signal can be accurately measured. Separation of the peaks is controlled by choice of column (see Chapter 6) and appropriate tactics (see Chapter 5), and detector performance and peak measurement and calculation are considered below.

8.2 DETECTOR

When a chromatographic detector gives a larger signal for what one knows to be a larger amount of component, it is both easy and reassuring to assume that the response is a linear function of concentration. In fact, this is generally true, but deviations from this simple rule of response do occur, and their significance depends upon the accuracy required. Each detector used should be evaluated for its response characteristics - the assumption that all detectors of a particular type have similar behaviour is false. Some general comments follow on the two most popular detectors, thermal conductivity and flame ionisation, and on the flame photometric detector which in the sulphur mode is known to deviate considerably from linearity. Deans (1968c) has described the nature and significance of detector characteristics, and a method of testing detector linearity is the subject of a British Standard currently in preparation.

8.2.1   Thermal Conductivity Detector

The response is a linear function of component concentration up to a certain
value, after which the incremental response per additional unit of concentration falls off.   Since the TCD is concentration-sensitive, a large sample
size can be accommodated by using a larger diameter column, with the
resulting higher flowrate of carrier gas.   Most modern TCDs employ a
constant filament temperature whereby the power supply to the bridge is
varied to allow for changes in thermal conductivity.   This system has a
greater linear range than the old type of constant voltage bridge.

8.2.2   Flame Ionisation Detector

Response depends upon mass flowrate of component into the detector, which
under fixed conditions translates to the more understandable component
concentration and sample size.   Most FIDs have some variation from linearity over their working range of 10^6:1, and this is frequently in the form
of an S-shaped response curve, response being lower than expected at some
points, and higher at others.   These variations can reach 5 or 10% relative
in some cases.   It is often possible to improve detector linearity by
adjusting the hydrogen and air flow rate to values other than those chosen
for optimum response.

8.2.3   Flame Photometric Detector

In the sulphur mode the response varies as the square of sulphur concentration, up to a limiting value imposed by the photomultiplier tube.   The
square root of the signal should be taken in each case before standard and
sample are quantitatively compared.   So long as peaks are compared one for
one, i.e. a component in the sample with the same component in the standard,
then the square root of either area or height can be used.   Comparison of
the square roots of areas of different peaks is incorrect since the square
root of area involves the square root of the time component, which is
different for different peaks.   If the FPD signal is treated in such a way
that the square root is extracted before area measurement, then the areas of
different peaks can be compared, as with a conventional detector.

If there is any sulphur background present in the detector, the response
becomes a complex function of concentration (see section 2.5.4), and
quantitative measurement becomes very difficult if not impossible.

8.3   SIGNIFICANCE OF NON-LINEAR RESPONSE

The section of the response curve which is of interest is that between the
concentrations of standard and sample, or over the range of component
concentrations to be expected.   In this region, the response may be assumed
to follow a straight line, of formula

$$y = a + bx$$

where x is the concentration and y is the response.   This is illustrated
schematically in Fig. 8.1.   When standard and sample are compared an error
is involved if the response is assumed to be linear through the origin.

The error can be expressed as

$$\text{Error} = \frac{a\,(x_{std} - x_{sample})}{a + bx_{std}}$$

Consequently the error tends to zero either as a tends to zero (i.e. the line does pass through the origin) or as the concentrations in sample and standard converge.

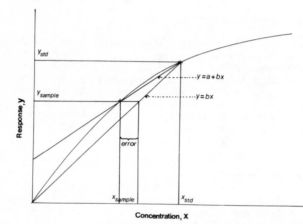

Fig. 8.1.   Error due to non-linear response

The error can be reduced by bracketing, where two different standard gases, with component concentrations on either side of that expected in the sample, are injected. If the concentrations in the two standards are x_1 and x_2, and the responses due to the standards and the samples are y_1, y_2 and y_s, then the sample concentration is given by

$$x_s = x_1 + \frac{(y_s - y_1)\,(x_2 - x_1)}{y_2 - y_1}$$

Bracketing is illustrated in Fig. 8.2.

8.4   PEAK AREA OR PEAK HEIGHT

Assuming that detector response can be plotted as a straight line passing through the origin, then the areas of peaks from mass-sensitive detectors, or from concentration-sensitive detectors used with constant carrier gas flow can be shown to be proportional to the masses of components. This is valid up to a concentration or sample size at which detector overload occurs.   Over a more limited range, peak width is independent of concentration, and so peak height is an empirical measure of mass.   Column overload, which causes peaks to become distorted and restricts peak height

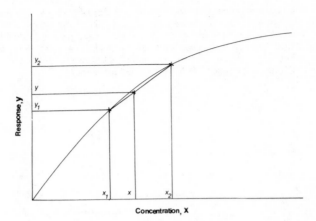

Fig. 8.2. Bracketing

measurement, usually occurs before detector overload in the case of the TCD or FID. With a more sensitive or selective detector, such as ECD, height or area should be equally valid within its working range.

Discussion on whether peak area or height should be preferred was common when measurement from the recorder chart was the usual method.   The recommended method of measuring area from the chart is as the product of peak height and width at half height.   Of the two, peak height measurement from the chart is recognised to be more precise.   The price and performance of modern integrators has changed the grounds for discussion, since they can measure either area or height with better precision and accuracy than any measurement from the chart.   Given this, any analyst using a modern integrator will almost invariably use areas, since they offer an advantage in linear range and allow peaks of different widths to be compared.   There may be occasional instances in which height would be preferred, as for example the case of overlapping peaks where height remains more constant than area with varying amounts of overlap.

8.5   INTEGRATORS

Integrators are described in section 2.7.   The most important features are:-

- wide dynamic range (10^6:1).
- ability to track baseline drift.
- area allocation for fused peaks by vertical drop or tangent skim.
- ability to set or force baselines where they may not be found automatically due to the complexity of the chromatogram.
- indication of how the integrator is making decisions.

Some integrators allow calibration at a number of levels, thus allowing for non- linear response.   The different calibration levels can be used to generate the equation of the curve, or the response can be assumed to follow a linear interpolation between adjacent calibration points.   There is little to choose between these methods since the more calibration levels that are used, the better the response curve will be defined, and the more closely the series of straight-line segments will approach it.   Figure 8.3 shows how such a curve may be constructed by an integrator.

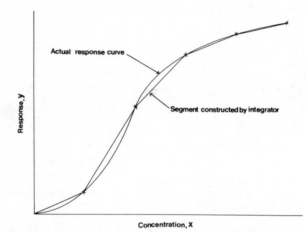

Fig. 8.3.   Multi-point calibration

The chromatographer need not understand the electronic techniques used by an integrator, but should appreciate how it decides on area allocation and baseline positioning, and how to change these decisions, if necessary, so as to coincide more closely with those which he or she would make.

Most integrators now include calculation facilities, to convert the area or height data into component concentration.   This is less important than raw area or height measurement, as it involves fairly straightforward mathematical transformation, but it can save a lot of time and potential transcription errors.

8.6   METHODS OF CALCULATION

The usual methods of calculation of concentrations from raw area or height data are internal standardisation, external standardisation and normalisation.   In very rare instances, such as operation of the ECD in the coulometric mode (section 2.5.3), the calculation can be direct.   The use of an internal standard, whereby a known amount of a component is added to the sample, is not practicable in gas analysis.   External standardisation, in which a sample and standard mixture are compared, is very popular: the gas sampling valve allows excellent repeatability of sample size, thus overcoming the major objection to this method when applied to liquids or solids.   Normalisation involves application of response factors to raw

areas, and summation of the corrected values to 100%.   Response factors can
be taken from the literature, but for accurate work should be established
for each application and regularly checked.   Since response factors are
established by injection of standard gas mixtures, normalisation can be
regarded as a modification to the external standard method for use in those
cases where all components in the gas sample are measured.

8.6.1   External Standard

A standard mixture of known composition is analysed under the same conditions as the sample.   Assuming linear response, and signals of y_{std}
and y_{sample}, then the concentration in the sample is given by

$$x_{sample} = \frac{x_{std} \cdot y_{sample}}{y_{std}}$$

If necessary, the bracketing technique (8.3) can be used.   In some
instances, it may not be practicable to have a component present in the
standard − it may for example be reactive or easily adsorbed if stored in a
cylinder.   In this case, a relative response factor, K, can be derived
which allows the component in the sample to be compared with a different
component in the standard.   The concentration is expressed as:

$$x_{sample} = \frac{K \cdot x'_{std} \cdot y_{sample}}{y'_{std}}$$

where x'_{std} and y'_{std} are the concentration and peak area of the
different component in the standard.   K can be derived by preparing a
dynamic standard containing both components (see Chapter 9), and should be
checked at regular intervals.

The external standard method is used when it is either not necessary or
practicable to analyse for all components in a mixture, or when the complete
analysis of a mixture involves more than one separation.   Its accuracy is
very dependent on repeatability of sample size.   The gas sampling valve
allows this, but the analyst must make sure that both standard and sample
reach the same conditions of temperature and pressure in the loop before
injection.   A further source of error is any difference in compressibility
factor between standard and sample − the sampling valve defines a volume of
gas but the number of moles can vary significantly with the composition of
the gas.   Thus a sample containing 1% molar nitrogen in butane contains 3%
more nitrogen in relative terms than a sample containing 1% molar nitrogen
in methane.   The detector, looking only at the nitrogen peak, sees this 3%
relative difference, and the chromatograph is in fact responding to the
number of moles per unit volume.   This is a further reason why the standard
and sample should be as similar as possible, both qualitatively and
quantitatively.

8.6.2   Measurement by Difference

Most analysts when asked to measure the purity of a gas will unhesitatingly
measure the impurities, and report the main component concentration by
difference, i.e. as 100% minus the sum of the impurities.   The same
analysts, however, would not measure a component at the 50% level by
difference, but directly.   Clearly there is a level above which measurement

by difference is preferred, because the sum of the absolute errors for the minor components is less than the absolute error for the major component when measured directly.   In the authors' experience this will be true for major components at 80% and higher and may be true for lower levels depending upon the nature and complexity of the sample.

8.6.3   Normalisation

If analysis accounts for all the components known to be present in a mixture, it should add up to exactly 100%.   The reasons why this rarely happens are non-linear response to one or more components, short-term changes in response which cause random variations in the ratios of peaks in the sample, and run to run changes such as a small variation in effective sample size between standard and sample or a step change in detector sensitivity.   These run to run changes affect all components in a sample equally, and so provided that the sum of all the components is close to 100%, it is valid to normalise to 100%.   Typically, an analyst might normalise if the components add up to between 99 and 101%, but look for the source of error if the variation from 100% is any greater.

The analyst should consider each case for normalisation on its merits. Factors which may influence the judgement are the advantages and drawbacks of the analytical method, the suitability of the standard with respect to the sample, the likelihood of undetermined components, and the units in which he or she is working.   Thus it is impossible that a gas should contain more than 100% molar, but as described in 8.6.1, the chromatograph responds to moles per unit volume, and so could apparently find more than 100% if the sample is considerably less ideal than the standard.

In the authors' laboratory, many natural gas samples have been analysed, and the methane contents of between 80 and 95% have been calculated directly using an external standard, directly with normalisation, and by difference. The results lead to the conclusions that at this level:-

(i)   Direct measurement is the least precise and probably the least accurate method.

(ii)   Measurement by difference is considerably more precise and usually more accurate.

(iii)   Direct measurement followed by normalisation is the most precise and accurate method.   Consideration of the arithmetic involved shows that it must produce a very similar result to measurement by difference.

CHAPTER 9

The Preparation of Standard Mixtures

9.1   INTRODUCTION

Gas chromatographs are, with rare exceptions, comparative and not absolute instruments.   When analysing gases, the sample introduction valve allows excellent repeatability, and so quantitative measurement is easily achieved by calibration with gas mixtures of known composition.   Although acceptable mixtures can be supplied from a number of commercial sources, there is a need for the analyst to prepare standard mixtures for himself if only to establish confidence in the commercial products.

9.2   DIRECT STANDARDS AND DERIVED STANDARDS

The adjectives primary and secondary are often applied to standard mixtures to indicate a position in a hierarchy based on confidence in the composition.   Thus, primary is variously used for standards prepared by a particular and well-regarded method (e.g. gravimetric), for standards issued by a particular supplier and whose use may be mandatory for legislative or regulatory purposes, or for standards prepared by a single-stage process. Secondary may imply that the preparation method is not the best, that the composition has been inferred from analysis, or that a two-stage process has been used, the primary standard having been diluted further.   The nomenclature is not satisfactory and we suggest that the terms direct and derived are more appropriate.   A direct standard is one whose composition is arrived at by consideration of the quantitative aspects of the method of preparation.   A derived standard is one whose composition is arrived at by analysis, often by comparison with a direct standard, and when appropriate it may be a sample of the gas to be analysed.   The methods described below produce direct standards.

9.3   ERRORS INVOLVED IN STANDARD PREPARATION

Three types of error can be defined:-

(i)   Aim, i.e.how close is the mixture to the desired composition.

(ii)   Repeatability, i.e. how consistent are the compositions of several mixtures prepared following the same technique.

(iii)   Accuracy, i.e how well is the final composition known.

The choice of method will be influenced by these factors in the context of the analytical requirement for which the standard mixture will be used. The choice will also be influenced by availability of equipment and time. In general, accuracy is the most important requirement for a standard mixture for chromatography.   If, however, a detector is known to have a non-linear response, then a better aim, so as to achieve greater similarity between standard and sample, may be preferable.

9.4   CATEGORIES OF METHODS

Methods fall into two general categories, static and dynamic.

9.4.1   Static Methods

These result in a single discrete body of standard mixture suitable for storage and transportation.   Examples described below involve adding known masses of gases (gravimetric method), adding known volumes of gases at constant pressure (volume increment method) or adding known pressures of gases at constant volume (pressure increment method).

9.4.2   Dynamic Methods

These involve the mixing of two or more gas flows or generation of a vapour into a gas flow.   The standard is usually prepared at the time and place at which it is needed and transportation and storage is difficult.   Dynamic methods are very valuable for standards containing trace components as with a flowing system "conditioning" of transfer lines can be more readily achieved.   Examples described below include the mixing of two monitored gas flows (proportioning pump method), addition of a minor component to a gas flow by injection (syringe pump method) and addition of a vapour to a gas flow by diffusion (permeation tube).

9.5   DESCRIPTION OF METHODS

The six methods described below are those more likely to be used when preparing standard mixtures for chromatography.   These and several alternatives are fully described in a series of ISO standards recently published or in preparation.   Table 9.1 compares the six methods for the various types of error.

9.5.1   Gravimetric Method

Gases are added successively to a high pressure cylinder, and the mass of each component calculated by weighing the cylinder before and after each addition (Daines, 1969, Szeri, 1974).   This is the most accurate method of preparing standard mixtures – although the masses of components are small by comparison with that of the cylinder, a good balance has a ratio of capacity to discrimination of $10^7:1$ and can weigh these small gas masses very accurately.

TABLE 9.1   Comparison of Methods of Preparation

Method	Aim	Repeatability	Accuracy
Gravimetric	1-5%	2%	0.1 - 0.5%
Volume increment	Certain values exactly	0.3%	1%
Pressure increment	1-5%	2%	1-10%
Proportioning pump	Certain values exactly	0.3%	1%
Syringe pump	Exact	1%	5%
Permeation	Exact	2%	5%

The amount of standard available depends upon the balance.   Thus, several litres can be prepared on a 1 kg balance which weighs to 0.1 mg or several cubic metres on a 100 kg balance weighing to 10 mg.

The volume of the cylinder is large relative to the incremental masses of gas, and so changes in atmospheric buoyancy such as would be caused by changes in temperature, barometric pressure and humidity can be significant at the level of accuracy obtainable. This can be compensated for by weighing against a "tare" cylinder, of similar size but on which no other operations are performed.   The cylinder to which components are added must be allowed to come to thermal equilibrium with its surroundings before the weighing will be reliable.   An alternative approach is to weigh under vacuum, thus eliminating atmospheric buoyancy.   This technique may not, in fact, save time, since evacuation can only slowly remove the adsorbed water layer on the cylinder surface, and constant weight may only be achieved after some time.

If the mass of a minor component is below a certain value, the method loses accuracy and preparation by double or even triple dilution should be considered.   Alternatively the minor component can be weighed into a small double-ended cylinder on a balance with better discrimination, and then fully displaced by purging with major component into a larger cylinder which has been weighed on a bigger balance.   The components are usually added by pressure increment (see below), and so this technique controls the aim and repeatability of the method.   Mixing of the components is not essential between weighings, but it is before further dilution or use.   The mass composition is converted to molar by use of accurately known molecular weights.   ISO 6142 (1981) treats some aspects of this in more detail.

9.5.2   Volume Increment Method

A component gas is isolated in a vessel of known volume, at atmospheric temperature and pressure and then completely displaced into a larger vessel, also of known volume.   The component can be displaced with mercury or with a piston, or by being purged out with the complementary gas.   The smaller vessel is a gas pipette or a gas-tight syringe, and the larger vessel is a flask or receiver with facilities for mixing the gases.   The larger vessel may be filled to slightly above atmospheric pressure, to facilitate removal of the mixture for analysis, in which case this excess pressure must be carefully measured.

The volumes of the vessels used can be accurately measured by weighing their water contents. The aim of the method depends upon the various sizes of the vessels available, and the composition which a particular pair of vessels will produce is known in advance. Repeatability is very good, and so is accuracy for components with near-ideal behaviour. When components with greater deviations from ideality are used, problems arise due to the fact that compressibility factors of mixtures are not linear interpolations of the individual component values. The following example, which is deliberately rather extreme, illustrates the point. Suppose a 50/50 nitrogen/n-butane mixture is wanted, and vessels of volumes x and 2x litres are to be used. The mixture can be made in three ways, each of which produces a different result:-

Method		Molar %
1.	Measure x litres of nitrogen, dilute to 2x with n-butane	N_2 49.47 $n-C_4$ 50.53
2.	Measure x litres of n-butane, dilute to 2x with nitrogen	N_2 48.67 $n-C_4$ 51.33
3.	Measure x litres of each, mix in larger vessel	N_2 49.03 $n-C_4$ 50.92

These variations in the composition are not large enough to cause the mixtures to have significantly different compressibility factors, and so the chromatograph would see these relative molar ratios.

Even at the largest practicable scale of operations, the amount of standard gas produced is small by comparison with other static methods. The mixtures are also less easily transported. ISO 6144 (1981) describes two techniques in more detail.

9.5.3   Pressure Increment Method

Components are added successively to a cylinder, with the pressure increments defining the quantity. The precision of this method depends more on the care and time taken than for most other methods; its aim for a one-off mixture is not very good, but the repeatability can be made good with care. The accuracy of the method depends upon whether the mixture is homogeneous and in thermal equilibrium when the pressure is measured - it can be good for simple mixtures of near-ideal gases.

Gases at high pressure mix much more slowly than is commonly realised. Hence, in addition to the temperature rise of the cylinder and its contents as components are added, which confuses the pressure readings, there is also a "piston" effect. This can cause error in two ways. If, for example, butane has been added, at a pressure lower than its saturation vapour pressure, and then another component added on top of it, the butane can be compressed into the lower part of the cylinder, where some of it may condense. This causes a large volume change, and hence an error in the pressure measurement. The second source of error concerns the compressibility factor for high-pressure mixtures. Deviations from ideal behaviour are much greater at high pressure (methane at 7 MPa (1050 psi) contains 15% more gas than predicted from ideal behaviour, and CO_2 at 3 MPa (450 psi) 22% more) and calculation of mixture compressibility is much more complex. Furthermore, a mixture will not have attained its correct compressibility factor until it is homogeneous, and so any pressure reading taken before this will not be correct.

In spite of these objections, the method is important as it allows large quantities of mixture to be prepared in a form most convenient for storage and manipulation. It is the normal method of preparing derived standards. ISO 6146 (1979) describes methods of calculation of composition from pressure.

9.5.4   Proportioning Pump Method

This is a widely used dynamic method. The precision pump has two pistons, one of which is driven at a constant speed by a synchronous motor, and the other via gear wheels at a fixed proportion of this speed, which can be changed by changing the gears. The pump is a very convenient way of producing a range of binary mixtures. More complex mixtures or other ranges can be obtained by operating pumps in series. Gases must be supplied to the pump at atmospheric pressure, and the outlet pressure must not be allowed to rise too far, lest the back-pressure affect the ratio.

The aim of the method is restricted by the fixed component ratios available, but the mixture composition is predictable. Repeatability is good, as is accuracy. For the best results, correction factors should be applied to allow for the efficiency of filling of the pistons, which depends upon the gas and the mixing ratio. Factors are supplied with the pump, but should ideally be checked with, for example, an accurate soap-film flowmeter.

9.5.5   Syringe Pump Method

The plunger of a syringe is driven at a constant rate, discharging its contents into a diluent gas flow. The contents may be a gas or a liquid which evaporates into the diluent gas steam. The rate of discharge of the syringe contents (the minor component) is measured from knowledge of the syringe capacity and timing the rate of travel of the plunger. The major component gas flow is measured by conventional means, such as gas meter, soap film flowmeter etc. If the minor component is a liquid, it must evaporate instantly on discharge - this can be assisted by touching the needle tip against some glass wool packing, which gives a large surface area for rapid evaporation. To produce mixtures with the minor component at very low levels, it may be more convenient to dilute it in a non-interfering solvent and to discharge this mixture from the syringe.

Since both flowrates can be controlled, the aim of the method is good. Repeatability is good and accuracy fair. It has been successfully used for trace levels of sulphur compounds, when the non-linear response of the flame photometric detector means that good aim can be preferable to good accuracy.

9.5.6   Permeation Method

This is a very good method for ultra-trace levels.   The pure minor component, which must be condensible at ambient temperature, is sealed as a
liquid into a plastic tube.   For as long as some liquid remains in the tube
the permeation rate is only a function of temperature.   The tube is contained in a thermostatted enclosure through which the diluent gas passes.
The rate of permeation can be measured from the change in mass of the tube
over a period of time, and the flowrate of diluent gas by conventional
means.   For best results the tube should be kept at the temperature of use
throughout its life and, if the component is easily oxidised or hydrolysed,
only dry, inert gas should be allowed to contact it, lest reaction products
build up in the pores, confusing the weighings and altering the rate of
permeation.

The aim of the method is good, the repeatability fair and the accuracy
relatively poor.   In fact, at these very low levels, 5% accuracy is
adequate for most purposes.   This is descrbed in ISO 6349 (1978).

              9.6   CHOICE OF METHOD

The six methods described above, while not exhaustive, cover most applications and each performs a different role.   It is difficult to advise on
which is the best method for any given problem but the following points
ought to be considered.

(a)   For trace analysis of vapours in gases the syringe pump and
             permeation tube methods are the choice;   the permeation tube
             method is best suited when a standard needs to be available over
             a long time (days or weeks).

(b)   If many different levels of the same components are required the
             proportioning pump is suitable.

(c)   If a large amount of standard is required, the pressure increment
             method is suitable.   Weighings however are needed to use it for
             an accurate direct standard.

(d)   The gravimetric method involves effort but provides the best
             direct standard for gas mixtures which are stable when compressed
             into cylinders.

(e)   The effort per standard mixture is high for the gravimetric and
             pressure increment methods, but as they can produce large amounts
             of mixture, the effort per analysis may be quite low.

APPENDIX

The Sampling of Gases

A.1 INTRODUCTION

This appendix is concerned with sampling from gas streams, which, except for atmospheric gases, involves sampling from pipework or chemical plant. It does not describe sampling of, for example, liquids containing dissolved gas components, as this was referred to in Chapter 7. Nor does it describe sampling of non-gaseous components of gas streams, such as liquid or solid particulates : these are adequately described elsewhere (Cornish, Jepson and Smurthwaite, 1981).

The sampling process involves some or all of a number of steps as follows:-

(a)   Removal of an aliquot (sample) from a bulk quantity of gas which is representative of that bulk.

(b)   Alteration of the physical state of the gas so that it can be manipulated.

(c)   Transfer or transportation of the sample so that it can be analysed at another place.

(d)   Storage of the sample for analysis at another time.

(e)   Transfer of the sample to the chromatograph.

The analyst is often only involved with the last of these stages. It is preferable that he or she should be involved, at least initially, with every stage of sampling, so as to avoid the performance of analyses which may be both a waste of time and misleading. Where this involvement is not the case, the analyst must always be aware that the implications of the analysis only relate to the sample "as received", and should make others aware of this where necessary.

A.2 REPRESENTATIVE NATURE OF SAMPLE

The composition of a gas mixture flowing through pipework or chemical plant is usually homogeneous in cross-section, due to the high rates of diffusion of gases and the commonly turbulent flow. Whether the composition is homogeneous in time depends upon the nature of the process from which the gas is produced, and the opportunities for mixing between the point of

AGC-E*

production and the point of sampling.

Static systems into which gases are added in a non-homogeneous manner achieve homogeneity with time due to diffusional mixing, at a rate which is inversely related to the total pressure in the system. Homogeneous gas mixtures will not alter in their composition (e.g. by "stratification") unless subjected to physical or chemical treatment which separates components.

A.2.1   Homogeneity in Section

If a sampling point is located close to a pipework tee where two gas streams are mixed, it may not allow representative sampling, and should be moved further downstream.   Another complication can arise from the "boundary layer". Even in a fully homogeneous, turbulently flowing gas stream, there is a static layer of gas on the pipe wall. Samples taken at a slow rate from the pipe wall may not represent the bulk of the gas flowing through the system. This can be solved by sampling at a high rate, so that the sample flow disturbs the boundary layer around the sampling point, or better, by using a probe which projects some way into the flowing gas stream. Use of such a probe which can be adjusted to different positions across the pipework diameter allows the composition to be surveyed and homogeneity to be assessed.

A.2.2   Homogeneity in Time

Production of natural gas from a reservoir gives a gas stream whose composition remains constant to within the limits of analysis. Some gas-making process are cyclical in nature, following a repeated cycle of some minutes.   Other processes, such as steel-making, are monitored by measuring gas compositions which change in tens of seconds. The time-scale of the process should always be considered when devising an analytical schedule.

The implications of the size of operations is also instructive. A natural gas transmission main can comfortably carry 30 million cubic metres per day. A chromatograph will require a sample of around 1 ml. Analysis once an hour gives a ratio of gas analysed : gas transmitted of $1:10^{12}$.   If the sampling and analysis process takes 10 minutes, the body of the gas from which the sample was taken will be several kilometres downstream by the time a result is available. It is fortunate that in this instance, homogeneity in time is very good.

The length of sampling line is also important in that it should be rapidly purged so that the sample is current rather than historical. If the gas stream is at pressure, it should, where possible, be depressurised close to the point of sampling, so that the line may be more rapidly purged with the expanded gas.

Rapid gas analysis has been described (Davison, 1970) as a way of following changing compositions, but applications have been few.   More commonly "period" samplers have been used so as to obtain an average composition. These have been described for low-pressure applications (Dudden, 1954) and also for high-pressure (Welker Engineering Company, 1981).   In either case the sample can be averaged by time or by flow. The alternative approach to cyclical processes, or one-off laboratory experiments, is to take a series of "spot" or "snap" samples at time intervals smaller than the time required for analysis, and to analyse them subsequently. A recent application (Fung and Channing, 1982) used a multi-position valve with 16 sample loops. Samples can be flushed through and then isolated in these loops successively, and stored for subsequent analysis. Both sample collection and analysis are readily automated.

It is worth emphasising that changes in composition in a gas mixture feeding into a pipeline are likely to be conserved. In the experiment described in Chapter 6, in which tracer was added to a natural gas pipeline, the dispersion of the tracer peak was about 2 minutes after 4 hours travel. This reflects the relatively low level of longitudinal mixing in such systems, which among other things contributes to the success of capillary columns.

A.3   SAMPLE TRANSFER

The conditions of temperature and pressure in the system from which a gas sample is required may be very different from those normally found in a laboratory or in the environment of a process analyser. Pressure can be maintained by sampling lines or sample containers of suitable quality, but reduction of pressure to atmospheric does not usually cause alteration of the state of the sample (the exception being retrograde behaviour of hydrocarbon mixtures mentioned in Chapter 6). Reduction in pressure is associated with Joule-Thompson cooling which is much more likely to affect the sample - it may be necessary to preheat the gas, or to heat the valve or regulator across which the pressure reduction occurs if the flow is high.

Reduction in temperature from that of the process to ambient is likely to be associated with alteration of the sample by condensation of components. Water is a common component of process streams, often at levels which make condensation inevitable at ambient temperature. This has two effects: overall composition will be altered by removal of most of the water, and the presence of liquid water may allow reactions which would not occur in the gas phase (reaction of H_2S and SO_2 is a case in point).

Trace heating of sampling lines up to and including the gas sampling valve on the chromatograph is possible, although frequently the resulting composition reflects the dewpoint at the lowest temperature along the line. Addition of a dry inert gas to the sample in a known ratio at the point of sampling is also possible. This dilution reduces the danger of condensation, but may also reduce the concentrations of sample components to below the detection limit.

Selective removal of water is claimed for "Perma Pure" dryers (Perma Pure Products Inc., leaflet PD-103). These are designed like a heat exchanger, but the tubes are made of a material which adsorbs and permeates water. The water is removed from the dryer by a counter-current purge flow of dry gas through the shell surrounding the tube bundle. Operating pressures are up to 670 kPa (100 psi) and temperatures to 90°C. The rate of permeation is a function of the difference between the partial pressure of water on either side of the tubes, and so is enhanced if the volumetric flowrate of purge gas is greater than that of sample. If the sample is available at pressure, a fraction of the dried product can be expanded and used as purge gas. The authors have no direct experience of this device and cannot comment on the permeation rates of other gases, but the principle is attractive.

A.4   SAMPLE CONTAINERS

These take many forms - samples have been received in, among other things, lengths of gas pipe and beer bottles. The most common, however, are single- or double-ended steel or aluminium cylinders, glass bulbs with stopcocks at each end, and plastic bags. The choice of container should be made primarily on the basis of maintaining sample integrity, but the practicability of sampling in particular locations may also affect it.

Metal containers have the advantage of being able to store pressurised samples, which make transfer into the chromatograph easier, and give a substantial stock of gas. They are suitable for transportation of samples, and long-term storage of most gases. (Mixtures containing carbon monoxide should not be stored in steel cylinders, since metal carbonyls are slowly formed. This not only alters the CO content, but creates a considerable hazard).

Glass sampling tubes are suitable for transportation, given appropriate precautions, and can be used for storage although their small capacity makes them less attractive. Trace levels of higher molecular weight components can be lost by solution in the stopcock grease, or cross-contamination may occur from some previous sample by the same mechanism. Samples are frequently transferred to the chromatograph by displacement with a fluid, although other techniques are described in Chapter 2. Continued storage of samples after addition of a displacing fluid is less likely to preserve sample integrity.

Plastic bags are very convenient - they can be folded when not is used, the amount of sample they contain is obvious, and they can be squeezed so as to displace the sample. It is difficult to remove the last traces of air or previous sample if the bag has only one connection, and our practice has been to purge the bag thoroughly with carrier gas and then normalise the analysis to 100%. Diffusion of sample component out of the bag and of air components in is the principle reason for not attempting long-term storage. "Tedlar" is one of the least adsorptive or permeable materials, and is widely used in the motor industry for exhaust gas sampling. Experience with the use of such bags for hydrocarbon mixtures shows that component concentrations vary little with storage times of up to three hours, but that overnight storage allows significant changes.

A.5   SAMPLING INTO THE CHROMATOGRAPH

The techniques and equipment used have been described in Chapter 2. There is rarely difficulty at this stage, provided that the pressure and temperature of the sample in the gas sampling valve have been allowed to reach reference conditions (commonly ambient) or are carefully measured. When a sample is displaced from a glass bulb, the liquid used is often slightly acidified water (to reduce the solubility of acid gases). Mercury was popular, and less likely to affect the gas composition, but most analysts now prefer to avoid it.

If a gas sample has been taken from a hot process stream into a cylinder, condensation of heavy components can occur between sampling and analysis. It this has happened, as may be shown for example by a dew point close to ambient, the cylinder should be warmed to the sampling temperature, and kept in this condition for several hours before analysis.

Contamination of the sample by air is possible at each stage of manipulation, including transfer into the chromatograph. For this reason, it is advisable to include oxygen and nitrogen measurement in any analytical scheme so that contamination can be recognised and, where appropriate, allowed for.

References

STANDARD METHODS

ASTM D1717-65   Analysis of commercial butane-butene mixtures and
                 isobutylene by gas chromatography
ASTM D1945-81. Analysis of natural gas by gas chromatography. Both above in
                 Annual Book of ASTM Standards. American Society for Testing
                 and Materials, Philadelphia.

BS 1756   Methods for the sampling and analysis of flue gases. Part 4
                 : 1977 Miscellaneous analyses.
BS 3048; 1958   Codes for the continuous sampling and automatic analysis of
                 flue gases. Indicators and recorders.
BS 3282; 1969   Glossary of Terms Relating to Gas Chromatography.
BS 4587; 1970   Recommendations for the selection of apparatus and
                 techniques for the analysis of gases by gas chromatography.
                 All above published by British Standards Institution,
                 London.

IP 194/74 (81) Butadiene purity and hydrocarbon impurities by GC.
IP 264/72 (79) Analysis of LPG and propylene concentrates by GC.
IP 344/80   Light hydrocarbons in stabilised crude oils by gas
                 chromatography.
                 All above in Institute of Petroleum Methods for Analysis and
                 Testing Part I. Heyden and Son, London.

IP 181/76   Sampling Petroleum Gases including Liquefied Petroleum
                 Gases. In Institute of Petroleum Methods for Analysis and
                 Testing Part IV. Heyden and Son, London.

ISO 6142-1981   Gas Analysis - Preparation of calibration gas mixtures-
                 weighing methods.
ISO 6144-1981   Gas Analysis - Preparation of calibration gas mixtures -
                 Static volumetric methods.
ISO 6146-1979   Gas Analysis - Preparation of calibration gas mixtures -
                 Manometric methods.
ISO 6349-1979   Gas Analysis - Preparation of calibration gas mixtures -
                 Permeation method.
ISO 6568   Natural Gas - Simple analysis by gas chromatography
                 All above published by ISO Central Secretariat, Geneva.

TRADE LITERATURE

Airco (1980).   Instant alert to hazardous arsine and phosphine.   Airco
    Industrial Gases, New Jersey.
Alltech Assoc. (1978).   Chromatography Products Catalogue No. 35 28-29.
    Altech Associates.
Applied Science (1979a).   Freons in Catalog 22 P.34.
Applied Science (1979b).   Amines in Catalog 22 P.37. Applied Science Europe
    B.V.
Carle (1975).   New solutions to some old problems in gas chromatography (by
    Johns & Berry) in Carle Current Peaks, Carle Instruments Inc.,
    California.
Carle - Carle Applications Note T397 - B   Carle Instruments Inc.,
    California.
Gow-Mac 117S. Sterilising Gas Mixture Gow-Mac Instrument Co., New Jersey.
Gow-Mac 123S.   The content of Silcon tetrachloride, $SiCl_4$, Trichloro
    silane SiH_2Cl_2 in a silane mixture.   Gow-Mac Instrument Co., New
    Jersey.

BOOKS AND JOURNALS

Allen J.D., J. Billingsley and J.T. Shaw (1974).   Evaluation of the
    measurements of oxides of nitrogen in combustion products by the
    chemiluminsence method. J. Inst. Fuel, (December 1974), 278-280.
Al-Thamir W.K., J.H. Purnell and R.J. Laub (1979). Dusted columns : An
    approach to enhancement of ᵍas-solid chromatography. J. Chrom., 176,
    232-236.
Al-Thamir W.K., J.H. Purnell and R.J. Laub (1980). Enhancement of ᵍas-solid
    chromatographic column performance by inert solid dilution.   J. Chrom.,
    188, 79-88.
Al-Thamir W.K., R.J. Laub and J.H. Purnell (1977).   Gas chromatographic
    separation of all C_1-C_5 hydrocarbons by multisubstrate gas-solid -
    liquid chromatography.   J. Chrom., 142, 3-14.
Andrawes F.F., R.S. Brazell and E.K. Gibson (1980).   Saturation region of
    helium ionisation detector for gas-solid and gas-liquid chromatography.
    Anal. Chem., 52, 891-896.
Andrawes F.F., T.B. Byers and E.K. Gibson (1981).   Effect of helium purity
    on the helium detector response in the saturation region of the detector
    field intensity.   J.Chrom., 205, 419-424.
Andrawes F.F. and E.K. Gibson (1978).   The effect of gaseous additives on
    the response of the helium ionisation detector.   Anal. Chem., 50,
    1146-1151.
Annino, R., J. Franko and H. Keller (1971).   Mixed carrier gases in
    chromatography - A source of error.   Anal. Chem., 43, 107-109.
Applebury T.E. and M.J. Schaer (1970).   Analysis of Kraft Pulpmill gases by
    process G.C.   J. of Air Pollution Control, 20, 83-86.
Aubeau R., J. LeRoy and L. Champeix (1965).   Influence of the degree of
    hydration of the absorbant on the chromatographic analysis of permanent
    gases.   J. Chrom., 19, 249-262.
Bean J.R. and R.E. White (1977).   Unheated external column and inlet
    modification of a GC. for PH_3 determinations.   Anal. Chem., 49,
    1468-1469.
Berry R. (1962).   Analysis of milli-microlitre quantities of permanent gas
    mixtures.   in M. van Swaay (Ed.) Gas Chromatography 1962. Butterworths,
    London. pp 321-334.
Beuerman D.R. and C.E. Meloan (1962).   Simultaneous determination of
    carbon and sulphur in organic compounds by gas chromatography.   Anal.
    Chem., 34, 319.

Blurton K.F. and J.R. Stetter (1978).   Sensitive electrochemical detector for gas chromatography.   J. Chrom., 155, 35-45.
Bombaugh K.J. (1963).   Improved efficiency in gas chromatography by molecular sieve flour.   Nature, 197, 1102.
Boreham G.R. and F.A. Marhoff (1960).   Fuel gas analysis : an apparatus incorporating a multi-cell thermal conductivity detector.   In R.P.W. Scott (Ed) Gas Chromatography 1960.   pp 412-421, Butterworths, London.
Brody S.S. and J.E. Chaney (1966).   Flame photometric detector. J. Gas Chrom., 4, 42-46.
Bros E. and J. Lasa (1979).   Concentration characteristics of the helium detector for gas chromatography.   J. Chrom., 174, 273.
Bros. E., J. Lasa and M. Kilarska (1974).   Application of a helium detector for determining small amounts of carbon dioxide, carbon monoxide, methane, carbonyl sulphide and hydrogen sulphide in helium by gas chromatography.   Chem. Anal.(Warsaw), 19, 1003-1011.
Bruner F., G.P. Cartoni and M. Possanzini (1969).   Separation of isotopic methanes by gas chromatography. Anal. Chem., 41, 1122.
Bruner F., P. Ciccioli and G. Bertoni (1976).   Analysis of sulphur in environmental samples with specific detection and selective columns. J. Chrom., 120, 200.
Bruner F., P. Ciccioli and F. Di Nardo (1975).   Further development in the determination of sulphur compounds in air by gas chromatography.   Anal. Chem., 47, 141.
Bruner F. and A. DiCorcia (1969).   The use of high efficiency packed columns for gas solid chromatography.   1.   The complete separation of $^{14}N_2-^{15}N_2$.   J. Chrom., 45, 304.
Brunnock J.V. and L.A. Luke (1968).   Rapid separation by carbon number and determination of naphthene and paraffin content of saturate petroleum distillate up to 185ºC.   Anal. Chem., 40, 2158.
Charron M. and M. Maman (1980).   An electrochemical detector for the chromatographic measurement of sulphur compounds in natural gas.   13th International Symposium on Chromatography, Cannes, 1980.   Note.   In preprint but not in published proceedings.
Cirendini S., J. Vermont, J.C. Gressin and C.L. Guillemin (1973).   Rapid gas chromatographic analysis on Spherosil.   J. Chrom.   84, 21.
Clay D.T. and S. Lynn (1975).   Pretreatment of molecular sieve 5A to eliminate tailing of NO.   Anal. Chem., 47, 1205-1206.
Clemons C.A., A.I. Coleman and B.E. Saltzman (1968).   Concentration and ultrasensitive chromatographic determination of SF_6 for application to meteorological tracing.   Environmental Science and Technology, 2, 551.
Cooper L.S., A.B. Densham, A.J. DeRose and B. Juren (1968).   Some aspects of the reception and transmission of North Sea gas.   Gas Council Research Communication GC150.
Cornish D.C., G. Jepson and M.J. Smurthwaite (1981).   Sampling Systems for Process Analysis.   Butterworths, 1981.
Coulson D. (1966).   Selective detection of nitrogen compounds in electrolytic conductivity gas chromatography.   J. Gas Chrom., 4, 285.
Craven J.S. and D.E. Clouser (1979).   A fresh design for thermal conductivity detector.   Presented at 30th Pittsburgh Conference on Analytical Chemistry and Applied Spectroscopy, 1978.
Cremer E. (1951).   Reported by A.I.M. Keulemans in Gas Chromatography, Reinhold Publishing Corporation, New York, 1959.
Daines M.E. (1969).   The preparation of standard gas mixtures by a gravimetric technique.   Chemistry and Industry. 1969, 1047.
David D.J. (1974).   Gas Chromatographic Detectors, Wiley-Interscience. New York, 1974.

Davis R.E. and R.A. Schreiber (1957).   Double column gas chromatography; analysis of non condensable and light hydrocarbon gases by a combined gas-liquid, gas-solid chromatograph.   Presented at Advances in gas chromatography, ACS Symposium, September 1957.

Davison E. (1970).   Rapid Analysis of Simple Gas Mixtures.   Chromatographia, 3, 43.

Deans D.R. (1965).   An improved technique for back flushing gas chromatographic columns.   J. Chrom., 18, 477-481.

Deans D.R. (1968a) Presented at discussion session in C.L.A. Harbourn (Ed) Gas Chromatography 1968, 447-451. Institute of Petroleum, 1969.

Deans D.R. (1968b).   A new technique for heart-cutting in gas chromatography.   Chromatographia, 1, 18-22.

Deans D.R.(1968c).   Accurate quantitative gas chromatographic analysis. Part I. Method of calculating results.   Chromatographia, 1, 187.

Deans D.R., M.T. Huckle and R.M. Peterson (1971).   A new column system for isothermal gas chromatography.   Chromatographia, 4, 279-285.

Deans D.R. and I. Scott (1973).   Gas Chromatographic columns with adjustable separation characteristics.   Anal. Chrom., 45, 1137-1141.

DiCorcia, A. and F. Bruner (1970).   The use of high-efficiency packed columns for gas solid chromatography II. The semi-preparative separation of isotopic mixtures.   J. Chrom., 49, 139.

DiCorcia A. and R. Samperi (1975).   Evaluation of modified graphitised carbon black for the analysis of light hydrocarbons.   J. Chrom., 107, 99-105.

Dietz R.N. (1968).   Gas chromatographic determination of NO on treated molecular sieve.   Anal. Chem., 40, 1576.

Driscoll J.N. (1977).   Evaluation of a new photoionisation detector for organic compounds.   J. Chrom., 134, 49-55.

Dudden W.R. (1954).   A new gas sampling apparatus.   Gas World ,140, 621.

Ellis J.F. and G. Iveson (1958).   The application of Gas Liquid Chromatography to the Analysis of volatile halogen and inter-halogen compounds.   In D.H. Desty (Ed.) Gas Chromatography 1958, Butterworths, London.

Fish A., N.H. Franklin and R.T. Pollard (1963).   Analysis of toxic gaseous combustion products.   J. Appl. Chem., 13, 506.

Fung D.P.C. and M.W. Channing (1982).   Application of Multiposition Valves for Gas Sampling and G.C. Analysis.   J. Chrom. Sci., 20, 188.

Gaglya V.G. (1979).   Determination of the impurities hydrogen, carbon monoxide and ethylene in air by means of a thermochemical pyroelectric detector.   Zh. Anal. Khim., 34, 405-407.

Genty C. and R. Schott (1970).   Quantitative Analysis for the Isotopes of Hydrogen - H_2, HD, HT, D_2, DT and T_2 - by Gas Chromatography. Anal. Chem., 42, 7-9.

Gibbons P.A. and K.A. Goode (1968).   Sulphur selective detector in natural gas chromatography.   Gas Journal, 336, 27-29.

Glover J. (1976).   RF discharge detector gives improved analysis of trace gases.   Control and Instrumentation, (May 1976) 31-33.

Goldan P.D., F.C. Fehsenfeld, W.C. Kuster, M.P. Phillips and R.E. Sievers (1980).   Vinyl chloride detection at sub ppb levels with a chemically sensitised electron capture detector.   Anal. Chem., 52, 1751-1754.

Goode K.A. (1970).   Gas chromatographic determination of sulphur compounds in North Sea natural gases by a flame-photometric detector.   J. Inst. Pet., 56, 33-41.

Goode K.A. (1977).   In situ fractionation - A simple technique for analysing complex mixtures using a routine gas chromatograph. Chromatographia, 10, 521.

Green S.A. and H. Pust (1957).   Use of silica gel and alumina in gas adsorption chromatography.   Anal. Chem., 59, 1055.

Grice H.W. and D.J. David (1967).   Performance and application of an
ultrasensitive detector for gas chromatography.   J. Chrom. Sci., 7,
239-247.
Guillemin C.L., M. Deleuil, S. Cirendini and J. Vermont (1971).   Spherosil
in modified gas-solid chromatography.   Anal. Chem., 43, 2015.
Guillemin C.L., M. LePage and A.J. deVries (1971).   Silica microbead
"Spherosil" in gas chromatography.   J. Chrom. Sci., 9, 470-486.
Hachenberg H. and A.P. Schmidt (1977).   Gas Chromatographic Headspace
Analysis. Heyden, 1977.
Halasz I. (1964).   Concentration and mass flow rate sensitive detectors in
gas chromatography. Anal. Chem., 36, 1428.
Halasz I. and E. Heine (1962).   Separation of low boiling hydrocarbons by
gas chromatography using packed capillary columns.   Nature, 194, 971.
Hendifar A.R. and M.R. Tirgan (1978).   Gas chromatographic studies of vinyl
chloride in air by catalytic hydrogenation to ethyl chloride.   J.
Chrom., 161, 119-125.
Hollis O.L.   (1966).   Separation of gaseous mixtures using porous
polyaromatic polymer beads.   Anal. Chem., 38, 309-316.
Isbell R.E.   (1963).   Determination of HCN and $(CN)_2$ by gas
chromatography.   Anal. Chem., 35, 255.
James A.T. and A.J.P. Martin (1952).   Gas-liquid partition chromatography
: the separation and microestimation of volatile fatty acids from formic
to dodecanoic acid.   Biochem. J., 50, 679.
Janak J. (1953).   Chromatographic semi-micro analysis of gases.   Chem.
Listy.,47, 464-467.
Janak J., M. Krejci and H. Dubsky   (1958).   Use of zeolites in gas
chromatography 1.   Separation of hydrogen, oxygen, nitrogen, carbon
monoxide and methane.   Chem. Listy., 52, 1099-1107.
Jeffery P.G. and P.J. Kipping (1972).   Gas analysis by gas chromatography.
Pergamon Press Ltd., Oxford.
Johns T. and E.A. Berry (1975).   New solution to some old problems in gas
chromatography.   In Carle Current Peaks, Carle Instruments Inc.,
California.
Kaiser R.E. (1970).   Carbon molecular sieve.   Chromatographia, 3, 38.
Karlsson B.M. (1966).   Heat treatment of molecular sieves for direct
separation of argon and oxygen at room temperature by gas
chromatography.   Anal. Chem., 38, 668-669.
Karmen A. (1964).   Specific detection of halogens and phosphorus by flame
ionisation.   Anal. Chem., 36, 1416.
Keulemans A.I.M.   (1959).   Gas Chromatography, Reinhold Publishing
Corporation, New York, 1959.
Kolb B. and P. Pospisil (1977).   A gas chromatographic assay for
quantitative analysis of volatiles in solid materials by discontinuous
gas extraction.   Chromatographia, 10, 705-711.
Kremer L. and L.D. Spicer (1973).   Gas chromatographic separation of
H_2S, COS and higher sulphur compounds with a single pass system.
Anal. Chem., 45, 1963-1964.
Krockenberger D., H. Lorkowski and L. Rohrschneider (1979).   A selective
G.C. column for trace analysis of vinyl chloride in air.
Chromatographia, 12, 787.
Lawson A. and H.G. McAdie (1970).   The gas chromatographic determination of
nitrogen oxides in air.   J.Chrom. Sci., 8, 731-734.
Leibrand R.J. (1967).   Atlas of gas analyses by gas chromatography.   J. Gas
Chrom., 5, 518-524.
Little J.N., W.A. Dark, P.W. Farlinger and K.J. Bombaugh (1970).   Gas
chromatographic packings with chemically-bonded stationary phases.
J.Chrom.Sci., 8, 674.

Littlewood A.B., T.C. Gibb and A.H. Anderson (1968).   Computer analysis of
    unresolved digitised chromatograms.   In C.L.A. Harbourn (Ed) Gas
    Chromatography 1968.   Institute of Petroleum, London, 1969.   pp
    297-318.
Lovelock J.E. (1971).   Atmospheric fluorine compounds as indicators of air
    movement. Nature, 230, 379.
Lovelock J.E. (1975).   Solute switching and detection by synchronous
    demodulation in gas chromatography. J. Chrom., 112, 29-36.
Lovelock J.E., K.W. Charlton and P.G. Simmonds (1969).   The palladium
    transmodulator : a new component for the gas chromatograph.   Anal.
    Chem., 41, 1048-1052.
Lovelock J.E., R.J. Maggs and E.R. Adlard (1971).   Gas phase coulometry by
    thermal electron attachment.   Anal. Chem., 43, 1962-1965.
Lovelock J.E., P.G. Simmonds and G.R. Shoemake (1971).   Rare gases of the
    atmosphere. Anal. Chem., 43, 1958-1961.
Lukac S. and J. Sevcik (1972a).   A discussion of the detection mechanism and
    the response character of the helium detector.   Chromatographia, 5,
    258-264.
Lukac S. and J. Sevcik (1972b).   A discussion of the detection mechanism and
    the response character of the helium detector II.   Chromatographia, 5,
    311-316.
Lund Thomsen E. and J.E. Lovelock (1976).   A continuous and immediate
    method for the detection of SF_6 and other tracer gases by electron
    capture in atmospheric diffusion experiments. Atmospheric Environment,
    10, 917.
Lysyl I. and P. Newton (1963).   Evaluation of gas chromatographic columns
    for the separation of fluorinated material. Anal. Chem., 35, 90-92.
Madison, J.J. (1958).   Analysis of Fixed and Condensable Gases by 2-stage
    G.C. Anal. Chem., 30, 1859
Maggs R.J., P.L. Joynes, A.J. Davies and J.E. Lovelock (1971).   The
    electron capture detector - a new mode of operation.   Anal.Chem., 43,
    1966-1971.
Mason D.M. and B.E. Eakin (1961).   Calculation of heating value and
    specific gravity of fuel gases. Research Bulletin No. 32, Institute of
    Gas Technology, Chicago.
McTaggart N.G., C.A. Miller and B. Pearce (1968).   Quantitative hydrocarbon
    gas analysis using alumina packed glass capillary columns.
    J.Inst.Pet., 54, 265.
Miller D.A. and E.P. Grimsrud (1979).   Correlation of electron capture
    response enhancements caused by oxygen with chemical structure for
    chlorinated hydrocarbons. Anal.Chem., 51, 851-859.
Miller R.J., S.D. Stearns and R.R. Freeman (1979).   The application of flow
    switching rotary values in two-dimensional high resolution gas
    chromatography.   Journal of High Resolution Chromatography and
    Chromatography Communications, 2, 55-62.
Moore W.R. and H.R. Ward (1968).   The separation of orthohydrogen and
    parahydrogen. J. Am. Chem. Soc., 80, 2909.
Noble F.W., K. Abel and P.W. Cook (1964).   Performance and characteristics
    of ultrasonic gas chromatograph effluent detector.   Anal.Chem., 36,
    1421.
Novak J. (1975).   Quantitative Analysis by Gas Chromatography, Marcel
    Dekker, New York.
Patterson P.L. (1978).   Comparison of quenching effects in single- and
    dual-flame photometric detectors. Anal.Chem., 50, 345-348.
Patterson P.L., R.L. Howe and A. Abu-Shumays (1978).   A dual flame
    photometric detector for sulphur and phosphorus compounds in gas
    chromatographic effluents. Anal.Chem., 50, 339-344.

Pauschmann H. (1964).   Gas chromatographic determination of hydrogen using
    helium as carrier gas.   Z.Anal.Chem., 203, 16-20.
Pearson E.D.   and P.J. Hines.   (1977).   Determination of H_2S, COS, CS_2
    and SO_2 in gases and hydrocarbon streams by GC/FPD.   Anal.Chem., 49,
    123-126.
Phillips M.P., R.E. Sievers, P.D. Goldan, W.C. Kuster and F.C. Fehsenfeld
    (1979).   Enhancement of electron capture detector sensitivity to
    non-electron-attaching compounds by addition of nitrous oxide to carrier
    gas.   Anal.Chem., 51, 1819-1825.
Price J.C., D.C. Fenimore, P.G. Simmonds and A. Zlatkis   (1968).   Design
    and operation of photoionisation detector for gas chromatography.
    Anal.Chem., 40, 541-547. Publishing Corporation, New York, 1959.
Prior F. (1947).   Thesis, Innsbruch 1947, cited by A.I.M. Keulemans in Gas
    Chromatography, Reinhold Publishing Corporation, New York, 1959.
Purcell J.E. and L.S. Ettre (1965).   Analysis of hydrogen with thermal
    conductivity detectors.   J. Gas Chrom., 3, 69-71.
Purer A., R.L. Kaplan and D.R. Smith (1969).   Separation of Ne isotopes by
    cryogenic chromatography.   J.Chem.Sci., 7, 504.
Ratcliffe D.B.   and B.H. Targett (1969).   A gas chromatographic
    determination of organo halogen impurities in dichlorofluoromethane.
    Analyst, 94, 1028
Ravey M. (1978).   Mixed bis lactams.   A highly polar stationary phase for
    gas chromatography of low molecular weight hydrocarbons.   J.Chrom.Sci.,
    16, 79.
Ray N.H. (1954).   Gas Chromatography II, The separation and analysis of gas
    mixtures by chromatographic methods.   J.Appl.Chem., 4, 82.
Rein H.T., M.E. Miville and A.H. Fainberg (1963).   Separation of oxygen and
    nitrogen by packed columns at room temperature.   Anal.Chem., 35, 1536.
Riederer M. and W. Sawodny (1979).   Gas chromatographic separation of the
    spin isomers of hydrogen at ambient temperature.   J.Chrom., 179,
    337-341.
Rosiek J., W. Gudowski and J. Lasa (1976).   Analytical parameters of
    photoionisation detector for gas chromatography.   Chem.Anal.(Warsaw),
    21, 1251-1260.
Rupprecht W.E. and T.R. Phillips (1969).   The utilisation of fuel-rich
    flames as sulphur detectors.   Anal.Chim.Acta, 47, 439.
Saha N.C., S.K. Jain and P.K. Dua (1978).   A generalised and easily
    adaptable gas chromatographic method for the analysis of gaseous
    hydrocarbons.   J.Chrom.Sci., 16, 323-328.
Saltzman B.E., A.I. Coleman and C.A. Clemons (1966).   Halogenated compounds
    as gaseous meteorological tracers.   Anal.Chem., 38, 753-758.
Scott C.G. (1959).   Alumina as a column packing in gas chromatography.
    J.Inst.Pet., 45, 118.
Sievers R.E., M.P. Phillips, R.M. Barkley, M.A. Wigner, M.J. Bollinger, R.S.
    Hutte and F.C. Fehsenfeld (1979).   Selective electron-capture
    sensitisation.   J.Chrom., 186, 3-14.
Simmonds P.G. (1978).   Direct determination of ambient carbon dioxide and
    nitrous oxide with a high-temperature ^{63}Ni electron capture
    detector.   J.Chrom., 166,593-598.
Simmonds P.G., A.J. Lovelock and J.E. Lovelock (1976).   Continuous and
    ultra-sensitive apparatus for the measurement of air-borne tracer
    substances.   J.Chrom., 126, 3-9.
Simmonds P.G., G.R. Shoemake, J.E. Lovelock and H.C. Lord (1972).
    Improvements in the determination of sulphur hexafluoride for use as a
    meteorological tracer.   Anal.Chem., 44, 860-863.
Smith C.J. and P.M. Chalk (1979).   Determination of nitrogenous gases
    evolved from soils in closed systems.   Analyst, 104, 538-544.

Smith K.A. and R.J. Dowdell (1973).   Gas chromatographic analysis of soil
   atmosphere. Automatic analysis. J. Chrom. Sci., 11,655-658.
Spears L. and N. Hackerman (1968).   Analysis of F_2, HF, NF_3,
   trans-N_2F_2, and N_2F_4 mixtures by gas chromatography.   J. Gas
   Chrom., 6, 392-393.
Sternberg J.C., W.S. Gallaway and D.G. Jones (1962).   The mechanism of
   response of flame ionisation detectors . In N. Brunner, J.E. Callen and
   M.D. Wise (Ed), Gas Chromatography 1961, Academic Press, New York. 1962,
   pp 231-267.
Stufkens J.S. and H.J. Bogaard (1975).   Rapid method for the determination
   of the composition of natural gas by gas chromatography. Anal.Chem.,
   47, 383-386.
Sullivan J.J. (1979).   Millionfold dynamic range on the electron capture
   detector. Presented at 30th Pittsburgh Conference on Analytical
   Chemistry and Applied Spectroscopy, 1979.
Szeri L.S. (1974).   Standard gas and vapour mixtures for chromatography.
   J.Appl.Chem. and Biotechnology, 24, 131-141.
Terry J.O. and J.H. Futrell (1965).   Three column, two detector gas
   chromatographic method for simultaneous analysis of mixtures of fixed
   gases and hydrocarbons. Anal. Chem., 37, 1165.
Thompson B. (1977).   Fundamentals of Gas Analysis by Gas Chromatography.
   Varian Associates Inc., California.
Thornsbury W.L. (1971).   Isothermal GC Separation of carbon dioxide, carbon
   oxysulphide, hydrogen sulphide, carbon disulphide and sulphur dioxide.
   Anal.Chem., 43, 452.
Trowell J.M. (1971).   Reaction of nitrogen dioxide with Porapak Q.
   J.Chrom.Sci., 9, 253-254.
van Deemter J.J., F.J. Zuiderweg and A. Klinkenberg (1956).   Longtitudinal
   diffusion and resistance to mass transfer as causes of non-ideality in
   chromatography. Chem.Eng.Sci., 5, 271.
Verdin A. (1973).   Gas Analysis Instrumentation. The Macmillan Press Ltd.,
   London.
Vermont J., J.C. Gressin and C.L. Guillemin (1973).   Rapid gas
   chromatographic analysis on spherosil.   J.Chrom., 84, 21.
Vinsjansen A. and K.E. Thrane (1978).   Gas chromatographic determination of
   phosphine in ambient air. Analyst, 103, 1195-1198.
Walker D.S. (1978). Gas chromatography of some sulphur containing gases in
   the v.p.m. level in air using Tenax gas chromatography. Analyst, 103,
   397-400
Weast R.C. (1977a). In Handbook of Chemistry and Physics, 57th Edition, CRC
   Press, Ohio. E2.
Weast R.C. (1977b).   In Handbook of Chemistry and Physics, 57th Edition,
   CRC Press, Ohio. F210.
Welker Engineering Company (1971).   US Patent 3945770.   UK Patent
   1481046.
Willis, D.E. (1978).   Column switching technique for gas chromatographic
   analysis. Anal.Chem., 50, 827-830.
Wittebrood R.T. (1972).   Comparisons between a TC detector with constant
   filament temperature and a conventional katharometer.   Chromatographia,
   5, 454-459.
Yamane, M. (1964).   Photoionisation detector for gas chromatography.
   J.Chrom., 14, 355-367.
Zoccolillo, L. and A. Liberti (1975).   Rapid gas chromatographic
   determination of SF_6 in air. J.Chrom., 108, 219.

Chapter 2

2.3
Roy C. and E. Chornet (1981).   A Low—Cost Device for Chromatographic
    Analysis of Gas Mixtures at Reduced Pressures. J. Chrom.Sci., 19,
    480.

Chih—An Hu J. (1982).   Comments on "A Low—Cost Device for Chromatographic
    Analysis of Gas Mixtures at Reduced Pressures. J.Chrom. Sci., 20,
    531.

2.5.1
Patterson P.L., R.A. Gatten, J. Kolar and C.Ontiveros (1982).   Improved
    Linear Response of the Thermal Conductivity Detector.   J. Chrom.Sci.,
    20, 27.

Chapter 3

3.1
Stark T.J. and P.A. Larson (1982).   Separation of C_1–C_5 Hydrocarbons on
    Cross—linked Methyl Silicone Fused Silica WCOT Columns. J.Chrom.Sci.,
    20, 341.

3.2.2
Guha O.K. and K.P. Mishra (1981).   Effect of pH on the gas chromatographic
    behaviour of silica gel. J.Chrom., 219, 101.

3.2.4
Kusz P., A. Andrysiak and J. Bobinski (1982).   Separation of Mixtures of
    Permanent Gases and Light Hydrocarbons on Spherical Carbon Molecular
    Sieves by Gas—Solid Chromatography. Chromatographia, 15, 297.

3.3
Castello G. and G. D'Amato (1981).   Characterization of Chromosorb Porous
    Polymer Bead Columns by Gas Chromatographic Retention Values of Light
    Hydrocarbons and Carbon Dioxide. J. Chrom., 212, 261.

Chapter 5

5.6
Squier D. and A. Hill (1982).   A Method for Determining Mixtures of
    Hydrocarbon and Inorganic Gases Using a Single Column and Programmed
    Cryogenic Temperature G.C. J.Chrom.Sci., 20, 429.

Chapter 6

6.2
Kuster W.C., P.D. Goldan and F.C. Fehsenfeld   (1981).   Controlled
    Environment Portable Gas Chromatograph for In—Situ Aircraft or
    Balloon—Borne Applications. J.Chrom., 205, 271.

Marenco A. and J—C. Delaunay (1981).   Automated Gas Chromatographic
    Determination of Atmospheric Carbon Monoxide at the Parts—per—Billion
    Level. Anal.Chem., 53, 567.

Verzele M., M. Verstappe and P. Sandra (1981).   Determination of traces of
    nitrogen and argon in oxygen by a simple gas chromatographic method.
    J.Chrom., 209, 455.

Chapter 6

6.4 and 6.5
Sood A. and R.B. Pannell (1982). Coal Liquefaction Product Gas Analysis
    with an Automated Gas Chromatographic. J.Chrom.Sci., 20, 39.

6.4 and 6.9.1 (see also 5.6 above)
Konig H. and M. Hermes (1981). Separation, Identification and Estimation of
    Propellant Gases and Solvents in Aerosol Products by Gas Chromatography
    (in German). Chromatographia, 14, 351.

6.6
Weiss R.F. (1981). Determination of Carbon Dioxide and Methane by Dual
    Catalyst Flame Ionisation Chromatography and Nitrous Oxide by Electron
    Capture Chromatography. J.Chrom.Sci., 19, 611.

6.7
Stein V.B. and R.S. Narang (1982). Determination of Mercaptans at
    Microgram-per-Cubic Meter Levels in Air by Gas Chromatography with
    Photoionisation Detector. Anal.Chem., 54, 991.

6.8
Hecker W.C. and A.T. Bell (1981). Gas Chromatographic Determination of
    Gases Formed in Catalytic Reduction of Nitric Oxide. Anal.Chem., 53,
    817.

6.8.2
Gates W., P. Zambri and J.N. Armor (1981). Simultaneous Analysis of Polar
    and Non-polar Compounds by Gas Chromatography : The Separation of O_2,
    N_2 and NH_3 using Parallel Columns. J.Chrom.Sci., 19, 183.

Chapter 8

8.4
Kipiniak W. (1981). A Basic Problem – The Measurement of Height and Area.
    J.Chrom.Sci., 19, 332.

Appendix A2

Kuessner A. (1982). Low Temperature Sampling Technique for the Determin-
    ation of Less Volatile Compounds in Gaseous Mixtures. Chromatographia,
    16, 207.

Author Index

Abel, K. 29
Abu-Shumays, A. 28
Adlard, E.R. 24
Allen, J.D. 96
Al-Thamir, W.J. 48, 49, 90
Anderson, A.H. 36
Andrawes, F.F. 31
Annino, R. 57
Applebury, T.E. 95
Aubeau, R. 39, 85

Barkley, R.M. See Sievers, R.E.
  25
Bean, J.R. 101
Berry, E.A. 33
Berry, R. 30
Bertoni, G. 101
Beuerman, D.R. 52
Billingsley, J. 96
Blurton, K.F. 32
Bogaard, H.J. 88
Bollinger, M.J. See Sievers, R.E.
  25
Bombaugh, K.J. 49, See also
  Little, J.N. 48
Boreham, G.R. 63
Brazell, R.S. 31
Brody, S.S. 26
Bros, E.E. 31
Bruner, F. 49, 101
Brunnock, J.V. 40
Byers, T.B. 31

Cartoni, G.P. 101
Chalk, P.M. 96
Champeix, L. 39, 85
Chaney, J.E. 26

Channing, M.W. 127
Charlton, K.W. 35, 83
Charron, M. 32
Ciccioli, P. 49
Cirendini, S. 47, See also
  Guillemin, C.L. 47
Clay, D.T. 40, 96.
Cleaver, B.A. 108
Clemons, C.A. 100
Clouser, D.E. 20
Coleman, A.T. 100
Cook, P.W. 29
Cooper, L.S. 86
Cornish, D.C. 125
Coulson, D. 97
Craven, J.S. 20
Cremer, E. 2

Daines, M.E. 120
Dark, W.A. See Little, J.N.
  48
David, D.J. 17, 29, 97
Davies, A.J. See Maggs, R.J. 25
Davis, R.E. 62
Davison, E. 126
Deans, D.R. 10, 40, 74, 75, 76
  78, 91, 112
DeLieuil, M. See Guillemin, C.L.
  47
Densham, A.B. See Cooper, L.S
  86
DeRose, A.J. See Cooper, L.S.
  86
deVries, A.J. 47
DiCorcia, A. 49, 91, 101
DiNardo, F. 49
Dietz, R.N. 40, 96

141

Subject Index

145